Television News, Politics and Young People

Television News, Politics and Young People

Generation Disconnected?

Mike Wayne, Julian Petley, Craig Murray and
Lesley Henderson

palgrave
macmillan

First published 2010 by
PALGRAVE MACMILLAN

Palgrave Macmillan in the UK is an imprint of Macmillan Publishers Limited, registered in England, company number 785998, of Houndmills, Basingstoke, Hampshire RG21 6XS.

Palgrave Macmillan in the US is a division of St Martin's Press LLC, 175 Fifth Avenue, New York, NY 10010.

Palgrave Macmillan is the global academic imprint of the above companies and has companies and representatives throughout the world.

Palgrave® and Macmillan® are registered trademarks in the United States, the United Kingdom, Europe and other countries.

ISBN: 978–0–230–21935–9 hardback

This book is printed on paper suitable for recycling and made from fully managed and sustained forest sources. Logging, pulping and manufacturing processes are expected to conform to the environmental regulations of the country of origin.

A catalogue record for this book is available from the British Library.

Library of Congress Cataloging-in-Publication Data

Television news, politics and young people : generation disconnected / Mike Wayne ... [et al.]
p. cm.
Summary: "This book is an exploration of the extent to which young people in the UK are disaffected with traditional politics, and particularly the role played by televisual representations of the political process. The authors look at how television represents young people themselves, and at how young people use new forms of media to inform themselves politically" – Provided by publisher.
ISBN 978–0–230–21935–9 (hardback)
1. Television broadcasting of news – Great Britain. 2. Youth – Political activity – Great Britain. 3. Television and youth – Great Britain. 4. Youth – Great Britain – Attitudes. 5. Television viewers – Great Britain. I. Wayne, Mike.

PN5124.T4T45 2010
323′.04208350941—dc22 2010002716

10 9 8 7 6 5 4 3 2 1
19 18 17 16 15 14 13 12 11 10

Printed and bound in Great Britain by
CPI Antony Rowe, Chippenham and Eastbourne

Contents

Tables

Figures

Illustrations

Acknowledgements

This book comes out of a larger project that was made possible by funded support from the Arts and Humanities Research Council. The authors would like to thank the many individuals who helped in some way with the research project, including all the young people who participated in the focus groups and the facilitators of those groups. We would also like to thank Mark Reid at the BFI for his help and all the broadcasters who agreed to be interviewed for the project. Our thanks also to Peter Keighron who conducted and transcribed the interviews. Thanks are due to David Buckingham, Cynthia Carter, Stephen Cushion, Máire Messenger Davies and Sonia Livingstone for their advice in researching Chapter 8.

Introduction

The problem

In 2001, against a background of low voter turnout, the growing gap between party-political and more 'single' issue-based politics and the growing consensus on policy among the three main political parties in the UK (Labour, Conservative, Liberal Democrat), the BBC began a series of research projects exploring public perceptions of politics and the relationship between such perceptions and the consumption of news. As an early report called *Beyond the Soundbite* put it:

> There are strong indications that those people who did not vote [in the national elections of 2001] are the same people who rarely watch or listen to any of our political coverage. (Kevill 2002: 1)

The report goes onto to note that young people

> are disenchanted with traditional institutions and Westminster seems increasingly outdated and irrelevant to them. And the younger they are, the stronger is the sense of disengagement and disillusion. The message for the BBC is that, like Westminster, we haven't kept in step with the issues that matter most to people, and so we too can appear less relevant. (Ibid.: 1)

One of the programmes that emerged from the policy review was *The Politics Show,* which replaced Sunday morning's more austere *On the Record* presented by the veteran broadcaster John Humphreys. However, whether putting *The Politics Show* presenter Jeremy Vine in a sweater, as the new programme did, sufficiently addressed the root causes of the problem of viewer disconnection is very much open to question. The strategy of taking some of the formality out of television news and political discussion programmes like *The Politics Show* is not necessarily a bad idea. Certainly, television news is a genre that has a certain rigidity to it that needs challenging. Here is Matt Barbet, one of the younger broadcasters on UK television recounting the time when he did a report without his tie.

> Let's be honest the necktie is such an outdated garment but I still have to wear one. I tried doing a 6.30 bulletin one morning without my tie, not because I forgot but because I just thought, right, I'm not going to wear a tie for this one. Probably about 30 seconds after the bulletin ended the editor phoned up. She wasn't angry but she just said, put your tie back on, there's a good lad. How much of a risk was I taking there? Not really much of a risk. Did it undermine my authority? I don't personally think so. But that, and this is the BBC we're talking about, is the lack of risk taking right there in evidence.

This is a revealing anecdote, telling us the extremely limited room there is for individual autonomy/innovation, the rapidity with which the institution polices itself and restores 'order' and the overall conservatism of the genre of television news. Television news certainly finds itself in a bind, torn between trying to make itself more appealing, more welcoming, more inclusive and not losing what it perceives to be its 'authority', as well as its older and more middle-class audience. But presentational and stylistic issues, while not unimportant, are probably secondary to the more substantive issue of content and overall approach and here the focus remains stubbornly on Westminster. There are historical as well as contemporary economic, institutional and ideological pressures that push broadcasters in the direction of the Westminster filter for political news, and we will explore some of these elsewhere. But the result is that broadcasters are interlocked with an institution which has declining interest and appeal for a sizable part of the public, especially, although not exclusively, young people. This book comes out of a research project funded by the Arts and Humanities Research Council. The project had three strands that generated evidence for its investigation into the apparent disconnection between television news and young people. The project interviewed 26 senior broadcasters involved in news production and explored with them a number of questions around the relationship of television news to younger audiences, the representation of young people on television news, the relationship of broadcasters with politicians and the role of new technology in disseminating television news. The interviews provide an important insight into the nature and direction of thinking within the broadcast industry around news provision. Drawing on their own experiences, the broadcasters can help us piece together the shifting trends and the historical evolution of styles and approaches which different channels have adopted at different times in the recent past. The broadcasters also provide us with evidence of institutional practices behind the scenes and behind the screens, which can otherwise be hard to perceive.

At the same time this evidence has to be approached critically. The broadcasters display both critical self-reflection *and* blind spots. They are perceptive but also creatures of institutions around which an awful lot of academic literature exists suggesting that broadcast news speaks from a very narrow social and political viewpoint. So we draw critically on this data throughout the book and the reader is invited to look through all the interviews at their leisure on the project website at: http://www.brunel.ac.uk/about/acad/sa/artsub/filmtv/tvnewsyoungpeople.

The second strand of the project involved content analysis of one month of television news coverage in 2006. Through this analysis we investigated the connection between news provision and young people's interest in and engagement with politics and public affairs. Among the various areas we focused on were how (and how well) young people are represented on the news, how 'politics' (in the narrow, Westminster sense) is reported, and the boundaries of political or ideological diversity across news formats. We also examined the broad news agenda across these channels, observing the most commonly reported story subjects and relating these to the interests and concerns of younger viewers. We further identified which social or professional groups received the most attention and the most opportunity to contribute to news discourse as 'news sources' in interviews, 'sound bites' and 'vox pops', paying particular attention to the ability of the public, or 'ordinary' citizens, to be seen and heard on the news. The level of access given to younger citizens specifically was a major concern here, for obvious reasons. The programmes we analysed were nightly broadcasts for each of the five terrestrial channels as well as the cable channel Sky News. We also investigated the children's news programme *Newsround* (BBC1).

The third strand of the project involved focus groups with young people. We wanted to generate our own evidence base drawn from young people themselves, of their relationship to television news. Along with broadcaster perceptions, and actual news content, audience reception of television news constitutes the third moment in the circuit of communication. To what extent do young people themselves feel connected or disconnected from television news and political stories? What are their impressions of how young people are represented on television news? What differences in terms of reception are there as the category of age intersects with other factors such as class and ethnicity? What do young people themselves think television could do to provide a news service that feels like it is more open to them? These are some of the questions we explored in this strand of the research. With these three strands of the research, we hoped to be able to explore the points of connection and disconnection between broadcasters' perceptions and understanding of what they are doing and the problem concerning audience disconnection, the news content itself, and young people's own perceptions and understandings of television news and how it does or does not speak to them. But the problems which UK television news faces in relation to its audience are not unique to the UK. A useful point of international comparison is with the situation in the US.

News consumption in the US

Looking at news consumption in the US provides some useful international comparison with trends in the UK. Researchers in the US have been worrying about the declining interest of young people in news about public affairs since at least the 1990s. Underpinning this emerging disconnect between young people and news consumption has been a disconnect with the realm of politics itself. Drawing on a National Election Study in 1994, Stephen Bennett found this disconnection across all age ranges, but under-30s were less inclined to vote than over-30s even though both groups shared similarly critical and negative attitudes towards politics and politicians. A large majority (62 per cent) of those polled reported that they never or only rarely discussed politics at all (Bennett 1997: 48, see also Eliasoph 1998). This indifference to politics naturally feeds through to an 'avoidance of exposure to mass media coverage of public affairs' (Bennett 1998: 535) with under-30s especially 'less likely to read, listen to, or watch political news stories' (540). Drawing on data from the Pew Research Center, Bennett notes that television network news specifically is only watched *regularly* by 22 per cent of 18–29 year olds, although (and this is a consistent finding) local television news does better (registering 51 per cent as a regular source of campaign news). However, young people are 'twice as likely as those over 30 to report following media coverage of entertainment news very closely' (ibid.: 538). Bennett finds that those who identify with a political party are more likely to have exposure to news media than those under-30s who do not. This strongly indicates that an initial orientation to news about public affairs is determined by how well the wider political culture is able to engage with its citizens. The more engaged they are with politics, the more engaged they are with news about public life and politics. With turnout in US Presidential elections hovering at around or just below 60 per cent, a sizable part of the eligible voters clearly do not feel engaged enough even to take a trip to the voting booth once every four years. More recent research from the Pew Research Centre shows that the 1990s trends that concerned scholars are continuing today. Analysis of data taken during the 2004 presidential campaign shows that network television news as a source for information about the elections is continuing to decline. Nevertheless, as in the UK, television remains the *main* source of news for Americans, with CNN and Fox each being the most popular individual sources for 20 per cent and 22 per cent of Americans respectively (Pew Research Centre 2004: 3). However, quite foreign to the UK situation, audiences for these main networks are strongly skewed along party lines with twice as many Republicans as Democrats watching Fox and twice as many Democrats as Republicans watching the other major networks (ibid.: 4). Within a picture of overall decline for network television news, there are some significant shifts in consumption within the medium. There were, for example, increases in popularity amongst young people for television news magazine programmes such as *60 Minutes* and *20/20*. In 2000, 18 per cent of under-30s cited such programmes as a regular source of political news, but in 2004 the figure was up to 26 per cent (ibid.: 10). Comedy shows such as Jon Stewart's *The Daily Show* and *Saturday Night Live* also gained two percentage points (6–8 per cent) from 2000 when Americans were asked *where* they got their news about the campaign. Young people in particular seemed appreciative of these comedy shows, with 21 per cent saying they regularly found out something they did not know about the elections from such sources. The Pew's research finding about the attractiveness of comedy programmes to young people as a source of news led to some breast beating within the American media. It was often assumed that viewers were going to such comedy sources *instead of* more traditional news sources, although in fact the evidence is that viewers of a sophisticated comedy programme such as *The Daily Show* actually require a high level of knowledge about events to make sense of the comedy (Young and Tisinger 2006: 116). In the UK, the longest running news satire series is, of course, the BBC's *Have I Got News For You*? However, perhaps the most significant combination of comedy, stunts and news or investigative journalism was the *Mark Thomas Comedy Product* (1996–2000) and *The Mark Thomas Product* (2001–2002) series which came to an end after disagreements with the broadcaster, Channel 4, over how far they would support Thomas' brand of direct action around corporate accountability and manslaughter. This highlights some of the parameters within which conventional news and informational programmes operate. Their supposed 'objectivity' masks a quiescent and uncritical relationship to power. In America, satirical comedy on politics has frequently been accused of fostering a disabling 'cynicism'. Lance Bennett, however, turns the charge around and suggests that the real disabling cynicism comes from the mainstream news providers who often know how political spin and the media themselves collude in offering simple scripted narratives that 'distort more complex realities... Stewart does not offer us cynicism for its own sake, but as a playful way to offer the kinds of insights that are not permitted in more serious news formats that slavishly cling to official accounts of events' (Bennett 2007: 279). In America, the declining audience for television news as a source of information about public affairs has been offset by the growth of the Internet. The Pew Research Centre found that the Internet was a regular source for campaign news in 2004 for 13 per cent of Americans, up from 9 per cent in 2000 (Pew Research Centre 2004: 1). However this growth is not driven by a desire for alternative news sources, but presumably by the convenience of 'on-demand' consumption. The most popular sources for Internet news are AOL, Yahoo, CNN and *The New York Times* (ibid.: 4).

In 2007, the Pew Research Centre did an analysis of celebrity news coverage and consumption. This is significant because one of the easiest strategies that broadcasters in the UK have adopted in an attempt to halt the ageing profile of their news audiences has been to cover more celebrity news. Yet the Pew found that 87 per cent of the public thinks there was too much reportage of celebrity on television news. For example, in the two days following the death of Anna Nicole Smith (the former *Playboy* model who married an octogenarian oil billionaire and who died in 2007) approximately 24 per cent of news from all sectors was devoted to this story. Cable news was the worst offender with approximately 50 per cent of its news devoted to Smith's death (Pew Research Centre 2007: 2). According to Pew, 'Public interest did not match the amount of coverage, and 61% of Americans said the story was over-covered' (ibid.: 2). Of course, it could be argued that what respondents say to researchers about celebrity news reflects what they think is socially acceptable rather than their own individual behaviour or tastes.

Interestingly, young people tended to blame the public more than the media for the amount of celebrity coverage while over-30s tended to blame the media more than the public. This might indicate amongst the under-30s a certain guilty investment in the stories, but clearly there are strong institutional dynamics driving broadcasters in this direction, quite independent of audience preferences. In a revealing illustration of the international flow of news stories, the Anna Nicole Smith story also played out on UK television news. One of our broadcaster interviewees, James Dagwell, from BBC *News 24*, refers specifically to this story:

> I think inevitably there are going to be older viewers thinking 'Why are we seeing a story about Anna Nicole Smith on the Six O'Clock news? I don't care who she is, I don't know who she is.' Well, that's one minute out of their news programme so you won't piss them off but I think it's right that we do those kind of stories because it's what people are talking about. I was quite surprised by the Anna Nicole Smith thing because News 24 did quite a lot on it the night they announced she died. I didn't think they would because I thought do people know who she is? But they ran it as breaking news and kept coming back to it which I think was a really good idea because there's this whole thing about water-cooler news and what people are going to be talking about the next day in the office and we should be doing that kind of stuff so long as it's not tittle-tattle and it's actually a news story.

Yet it is hard to argue that this was in response to 'demand' from UK audiences for information on this specific story. Even Mr Dagwell seems unsure that Anna Nicole Smith was much of a known personality amongst the British public at the time. Rather, the story says more about the default orientations of UK television news towards celebrity themes when it tries to speak to a broader and younger audience.

News consumption in the UK

In 2002, the Independent Television Commission (the then regulator for television) and the Broadcasting Standards Council (the then monitoring and advisory body to broadcasters) published a report called *New News, Old News* by Ian Hargreaves and Jamie Thomas. The picture the report painted was of a somewhat contradictory situation in terms of the relationship between news and television audiences, but there were already some worrying trends. There were definite signs that audience engagement with the world of politics and public affairs, at least as it was being presented on television, was on the wane. Regular current affairs viewers had declined to only 16 per cent of the audience (Hargreaves and Thomas 2002: 6) while the 2001 General Election programme on the BBC was watched by only 4.9 million people, compared with 11.8 million in 1992 (ibid.: 9). Since 1994, total audiences for evening news bulletins were down 10 per cent (ibid.: 10). On the other hand television news remained the main source of news for 65 per cent of the audience (ibid.: 45) while 68 per cent of people turn on the television news at least once a day (ibid.: 47). The Internet, in contrast, was regarded as the main source of news by 3 per cent of 16–34-year-olds (ibid.: 46). There was a similarly contradictory picture in terms of audience satisfaction. For example, among 16–34-year-olds, 96 per cent were either 'very' or 'fairly' satisfied with their news provision (ibid.: 49). They also found high satisfaction levels when asked if television does a good job of informing them about UK politics, where only 7 per cent disagreed that it did (ibid.: 56). Yet when respondents were asked slightly different questions, another picture emerged. When audiences were asked how much they knew about UK politics, the figures were less encouraging: only 35 per cent felt averagely knowledgeable, with approximately 35 per cent knowing little or nothing (ibid.: 58). Amongst 16–34-year-olds, a striking 48 per cent said they knew little or nothing about UK politics (ibid.: 59). Of course, such responses are not necessarily the sole fault of television news, but if television news is the main source of news about politics, then it must shoulder some of the blame for this lack of confidence amongst viewers. This is reinforced by responses to other questions that elicited a somewhat less positive response to television news. The authors found that 25 per cent agreed with the statement that 'TV news is not as good as it should be' while a further 43 per cent were 'neutral' on this question. They also picked up evidence that viewers (34 per cent) felt that the issues that television news deals with feel remote and not 'relevant to me personally' (ibid.: 78). The authors' conclusions were similarly contradictory. On the one hand they concluded that there was 'no sign of dissatisfaction among young people with the choice of broadcast news programmes available to them' (ibid.: 83). Yet on the other hand they concluded that

> Television news is increasingly failing to connect with people from the C2DE class base, younger age base, and is not doing a good enough job of responding to concerns of citizens in certain

geographic regions of the UK and certain ethnic minority communities. (Ibid.: 80)

In 2003, the Office of Communications, OFCOM, took over as the regulator of broadcasting and telecommunications. It has conducted further research into the relationship between television news and audiences and is an important source of statistical evidence. The negative trends picked up in the Hargreaves and Thomas report appear to be continuing. OFCOM research has found that amongst 16–24-year-olds, viewing of network terrestrial news has decreased over time, from 41 hours a year in 2002, to 33 hours in 2006. Much of this decline was driven by ITV chopping and changing the time of its late evening news bulletin (OFCOM 2007c: 48). Nevertheless, OFCOM also finds that television remains the main provider of news for audiences (65 per cent), with newspapers trailing behind at 14 per cent and the Internet, although growing, a main source of news for only 6 per cent (2007a: 17). Amongst the television channels, the BBC dominates, taking 50 per cent of the audience share for television news. ITV has 26.8 per cent, Channel 4 4.5 per cent and Sky News 4.9 per cent (ibid.: 18). Once again a contradictory picture emerges. While television news remains the main source of news for most people there are strong indicators of public dissatisfaction with news provision. A survey statement that 'much of the news on TV is not relevant to me' was agreed by 55 per cent of people polled, up from the 34 per cent recorded by the Hargreaves and Thomas report (ibid.: 60). OFCOM recorded that viewers wanted to see more 'ordinary people' on the news (ibid.). Different parts of the audience felt aggrieved by aspects of their representation on the news. Black viewers complained that they were frequently linked to crime on television news while Muslims complained of being linked to extremism. At the same time that television news has these issues with its audiences, it is also under pressure from economic imperatives. News is an expensive operation to mount. For the BBC, by far the largest provider, costs for news provision were more than £112 million in 2005/2006 (ibid.: 20). Meanwhile the economic value of ITN's news contract to provide news for ITV and Channel 4 has been declining. Only Channel 4's public service obligations keep the hour-long *Channel 4 News* on at a peak time in its schedule. As Martin Fewell, deputy editor of *Channel 4 News* puts it:

We have a long-term contract with Channel 4 which goes until 2010 which means that the amount we get paid for producing the news is

> agreed five years in advance. So that insulates us from the vagaries of the economic cycle. It doesn't insulate Channel 4 from it but it means that we can invest in people and we can invest in equipment with a degree of certainty about what our revenue will be in that period. Secondly, Channel 4 has always made clear that the news is not about ratings. It's not about the number of adverts it can sell within the news. It's also said, as part of its ongoing discussions with Ofcom, that in effect it loses money on the news. In a way that's fairly obvious. It could run a much cheaper programme in that slot and make a lot more revenue. By scheduling your news at 7pm you are making a very conscious decision that you are not going to chase audiences in that period and therefore you are accepting that you are going to lose revenue. But, for all the reasons that we know, the news is absolutely at the core of Channel 4's public service commitment.

Public service commitments are essential to the survival of television news in its current form. When Channel Five was being set up in the mid-1990s, 'it was said that none of the consortia interested in bidding for the license wanted to carry any news programming at all. Eventually, news was mandated by Parliament as a PSB [public service broadcasting] requirement' (OFCOM 2007a: 23). Despite this, the overwhelming direction of broadcasting policy has been steadily increasing market pressures, imperatives and competition into the system since the 1990s. For example, the Broadcasting Act 1990 formed a new 'light touch' regulator the Independent Television Commission (meaning that ITV and the new Channel Five would have fewer public service obligations). ITV was opened up to foreign ownership and monopoly consolidation (formerly one company could only own one franchise). The Act also shifted Channel 4 into more commercial territory by forcing it to sell its own advertising to fund itself, thus placing it in direct competition for revenues with ITV and Channel Five. The Communications Act of 2003 established the openly pro-market regulatory body, OFCOM, to replace the already 'light touch' Independent Television Commission and further lifted restrictions on cross-media ownership. The results of this steady marketisation of broadcasting are apparent everywhere. ITV is pressing the communications regulator OFCOM to agree to reduce its obligations to broadcast children's programmes in the afternoon and has already closed down its production of homegrown children's programming. ITV is also looking to cut back on its local and regional news programming, while the value of ITN's contract to provide ITV's news is also falling (ibid.: 22). There is a looming financial crisis for Channel 4 and the BBC has been forced into another round of cuts following a disappointing settlement fixing the cost of the license fee until 2012. These cuts will hit news and current affairs programming particularly hard. In another sign of the times, the BBC shut down its online education provision, called BBC Jam because of pressure and complaints from private sector software suppliers that the BBC was damaging their interests. Meanwhile, trust in television was damaged by a series of premium rate telephone scandals in 2007. All of these problems are directly traceable to the increasingly prevalent market pressures being introduced into the system which are eroding the public service ethos of television in favour of a scramble for ever higher profits.

Conclusion

A lot is at stake in how one interprets the evidence and understands the causes behind public disengagement with both politics and news about politics, as well as the role of market forces within broadcasting. Such interpretations can inform public policy and broadcaster practice. It is significant in this regard that the regulator OFCOM covers both broadcasting *and* telecommunications. The world of telecommunications (telephones, Internet and cable) is very much the world of the market, of private companies operating to maximise profits and sell to consumers whatever goods and services that can meet that objective within any given set of conditions (conditions shaped to a large extent by the private companies and the imperatives of the market). The joining of broadcasting, which has traditionally had strong public service principles underpinning it, to the world of telecommunications is significant. OFCOM seems to be in the business of integrating television more fully into the universe of the market than it already is.

> [OFCOM's] leading figures – drawn largely from the worlds of advertising, cable, consultancy and politics – appear to have little interest in the qualitative dimensions of an audiovisual culture. Its ethos is predominantly neoliberal, and its language and organizing concepts are suitable for an analysis of markets and of competition, but not of social significance and cultural value. (Harvey 2006: 93)

Ultimately, it is only by asking what the social significance and cultural value of television news is, that we can answer the question of why television news matters. News matters because it offers representations about the world of public affairs and political action. Within such representations a variety of social actors appear and a variety of voices and perspectives are heard. But how diverse is that variety? Within what parameters do the basic founding assumptions of news operate in terms of what can be done, what can be said, what is 'realistic'? To what extent do the diverse formats of television news correlate with an actual diversity of views and perspectives? To what extent does television news operate independently of political and economic power in society at large? How diverse are the audiences that news even tries to address? Does television news imagine its audiences as active citizens, or as passive consumers of easily digestible news items? When OFCOM asks 'How can news help citizens of the 21st century to play a full part in a modern democracy?' (OFCOM 2007a: 13), we would have to ask some of the above questions relating to news to answer the question. These, however, are not the sort of questions that the regulator will ask nor indeed even broadcasting institutions such as the BBC in its public documents. While it will speak the language of citizenship and culture more confidently than OFCOM, such as in its document *Building Public Value*, there is little discussion of the structures of political, economic and social power in the UK and what implications these have for news provision and news consumption. The first two chapters in this book will seek to address that absence and will contextualise the issue of television news within an analysis of precisely those power structures. This is important if we are to understand how the public relate to the world of public affairs, how television news relates to these structures of power and how both audiences and news relate to each other in these contexts. Subsequent chapters will explore the institutional practices of television news broadcasters, exploring the continuities and differences between different providers. We shall explore the representation of young people in television news, the representation of politics in television news and the representation of political activity that takes place outside the mainstream political parties. In the final two chapters, we assess to what extent young people are making use of both old and new media in their consumption of news, and what young people themselves think of television news provision as a source of information about what is happening in political and public affairs.

1
The Crisis of Political Representation

To begin to understand why the realm of public affairs which television news reports on has a declining grip on the popular imagination, we need to situate television news within a much wider context than television journalism itself. We have to begin with the whole fabric of our political culture and what has been happening to it in recent decades if we are to understand why news about public affairs seems increasingly disconnected from the lives of various groups within society, including young people. For it ought to be clear that the crisis around watching news about public affairs is connected with the crisis around public and political life itself.

Political alienation in the UK

A good place to begin thinking about the crisis in politics and its relationship to television news and information programming, in the UK, is with the report by the Power Commission in March 2006. This report, funded by the Joseph Rowntree Foundation, was set up to explore the widespread disengagement from formal democratic politics, especially as it is incarnated in the Palace of Westminster. One manifestation of this disengagement has been the declining participation in national elections in recent years. As the winning party in 2005, New Labour polled 9.5 million votes. But 17 million registered voters failed to attend a polling station (Power Report 2006: 33). Such disengagement, the report notes, will lead to a crisis of democratic legitimacy, growing political exclusion, a breakdown in social consensus, a rise in undemocratic political forces and growing authoritarianism of government (ibid.: 15). To take just one indicator of the threat from the rise of anti-democratic forces, we can note that the British National Party, a fascist and white supremacist organisation, won two seats from British voters in the European elections of 2009. This was on the back of a very low voter turnout within the UK of 34.7 per cent.

The declining participation in national elections had become evident in 2001. Since 1922, the percentage of the electorate voting in national elections has been in the 70 and sometimes in the 80 per cent range (Foot 2005: 424). But in 2001, which saw the second re-election of a New Labour government, there was a sizable collapse in those figures, down to 59.4 per cent. The 2005 general election recorded only the slightest of recoveries in national participation, up to 61 per cent. In 2001 the *British Social Attitudes* survey found a 'clear change of mood' amongst the public as a whole vis-à-vis politicians in the 1990s. They recorded a declining trust in politicians and a declining sense of the political system being responsive to public views (Park et al. 2001: 203–5). This was not reversed by the election of New Labour in 1997 and so this trend was not exclusively about the long hold of the Conservative Party in office (from 1979, with the election of Margaret Thatcher, through to 1997, when John Major's Conservative government lost power). Thus the survey, conducted by cautious social scientists, concludes that there 'appears instead that there is a more fundamental crisis of confidence in the political system' (Park et al. 2001: 207). A few year later, as the trends picked up by the *British Social Attitudes* survey appeared to become entrenched, the Power report could put it more strongly: 'The current way of doing politics is killing politics' (Power Report 2006: 21). The report makes three main recommendations, two of which directly pertain to political institutions: first, a rebalancing of power away from the Executive (the Prime Minister and cabinet) and unaccountable bodies towards Parliament and local government; and second, electoral reform away from the first-past-the post system which currently means that a relatively small number of voters in 'swing' constituencies determine the outcome of national elections. Third, the report calls for greater citizen involvement and participation in political decisions and policy-making. This third recommendation shifts the focus away from political representatives, especially in Westminster, and instead recognises the importance of ordinary people becoming involved in and knowledgeable about politics. The report gives a special place to the role of broadcasters in encouraging a greater culture of 'political engagement' (ibid.: 121) and suggests that a

> ... requirement should be introduced that public service broadcasters develop strategies to involve viewers in deliberation on matters of

> public importance – this would be aided by the use of digital technology. (Ibid.: 24)

There can be little doubt that electoral reform would help open up the political process to new voices and generate a public debate about policy choices and options. As new electoral systems have been developed in Scotland and Northern Ireland around devolved assemblies, it seems feasible that they may one day be used in relation to the election of MPs for Parliament. It would also be highly desirable if institutional changes making the Executive more accountable to Parliament could be implemented. Yet, here we may come up against structural obstacles to accountability that make such changes highly unlikely outside contexts of radical social change. Likewise, while the emphasis on the important role public service broadcasting *could* play in encouraging a culture of political engagement is welcome, there needs to be a hard-headed understanding of the institutional problems and difficulties obstructing just such possibilities. Ironically, for a report called Power, there is in fact very little coherent analysis of the way power is structured in Britain. The report is at its weakest where it comes to linking its diagnosis of the problems around the widespread 'disconnect' between political representatives and the public, with the causal factors responsible for the current situation. For the Power report, the main problem appears to be that politics has in some way failed to keep up with the times, failed to be modern enough to change and be responsive to the 'diverse and complex values and interests of the individuals which make up our post-industrial society' (ibid.: 19). This is an inadequate formulation of the problem as we shall see, although no more inadequate than the responses to the crisis that come intermittently from the political establishment itself, which now resembles, in its isolation from public opinion and indeed contempt of ordinary people, the court of Louis XVI before the French revolution.

We need to situate both the system of political representation, known as representative democracy, *and* television news and information programming, within a larger context and structure of power, especially socio-economic power. And before we get to *television* news we need to take a longer historical perspective on the rise of the news media in relation to the press. We need to excavate the philosophical principles underpinning the 'free press' and the close relationship that was forged between the production, dissemination and exchange of information, with notions around democracy. And we need to see how the media and democracy interacted with capitalism as it evolved over time.

Democracy, capitalism and the media

The classic philosophical account of the historical emergence and decline of the public sphere was developed by Jürgen Habermas in his book *The Structural Transformation of the Public Sphere*. As Habermas tells it, the initial seventeenth-century development of a public forum for debate and the exchange of views, information, and arguments, broke the private power of the aristocracy, whose 'aura of feudal authority' (1996: 8) had dominated what public forums had existed within medieval Europe. The geographical and quantitative expansion of trade and a rising mercantile class required the development of means of communications, such as the mail (ibid.: 16). If news was initially restricted to information circulating amongst the new mercantile and administrative elites, the potential for news itself to become a commodity and to be sold to a broader public was the logical next step (ibid.: 21). Habermas contends (following Marx) that the projection of particular (property owning) interests as 'universal' interests (of benefit to the majority) had a kernel of truth for a *limited historical period* in the context of the emancipation of civil society from 'absolutist regimentation' (ibid.: 56). That emancipation from nonarchical absolutism began in earnest during the English Civil War (1642–1651). Although the Republican forces were victorious they were by no means internally unified. The Republican leader Cromwell and the large landed gentry were fighting for independence from the arbitrary and unaccountable rule of the monarch. They envisaged and implemented a constitutional oligarchy to replace feudal absolutism. Political control would pass to the landed gentry whose representatives would sit in a Parliament elected by the large landed gentry. Smaller property owners and the non-property owning commoners who also made up the Republican forces were drawn to far more radical and genuinely democratic currents (the Levellers, the Diggers, later the Fifth Monarchists) that argued for real effective equality between men (Manning 1999). These currents were marginalised and even crushed physically. Between the English Civil War and the late eighteenth century, 'democracy' was a dirty word for the property-owning classes. It was understood in its more or less original sense, derived from ancient Greece, as rule by the people (demos) and as such was to be avoided at all costs (Wood 1996: 231). Yet at the same time, the spread of market relations, of buying and selling between people who were formerly 'free' to enter into business relations with each other, including relations between large property owners and workers selling their labour to them, opened up a space for ideas to spread that went beyond the limited political freedoms won *for* the oligarchy of landed gentry and, later, industrialists. In Britain, by the 1730s, 'the press was for the first time established as a genuinely critical organ of a public engaged in critical political debate' (Habermas 1996: 60). The term 'public opinion', as a body of ideas formed by critical debate in the public sphere and as a force which the state had to recognise, emerged in the late eighteenth century in Britain (ibid.: 95). The public sphere thus emerged as a site of struggle, marking a shift in the nature of power from outright coercion to an increasing stress on consent; from the arbitrary will of the state in particular, to 'reasoned' use. Crucially, the public sphere provided the philosophical grounding and practical support for the extension of the popular franchise, which still nevertheless had to be struggled for and won *from below* by the Chartists and Suffragettes among others. The establishment of the popular franchise has profound *moral* effects within any society. It provides a resource with which to challenge authority in all its forms, it encourages the politicisation of apparently non-political social relations (thus opening them up for modification and democratisation) and it encourages, theoretically at least, diversity of political parties/organisations/policies 'within a frame of settled commitments' (Kateb 1981: 360–1). We shall see how the 'frame of settled commitments' has narrowed in recent years. But from the start, once we factor in the *economic-social* interests within which elections take place (precisely the context which liberals like to make disappear from view) then we can see how a basic problem of political communication is constitutive of a class divided society that has regular elections. Eric Hobsbawm has described how the open contempt for democracy amongst the elites gave way to hypocrisy, duplicity and manipulation as mass electoralism advanced in the late nineteenth and early twentieth centuries.

> What candidate wanted to tell his voters that he considered them too stupid and ignorant to know what was best in politics, and that their demands were as absurd as they were dangerous to the future of the country. What statesman, surrounded by reporters carrying his words to the remotest corner tavern, would actually say what he meant? (Hobsbawm 1987: 87)

We have seen so far then that the relationship between capitalism and democracy is paradoxical and complex. On the one hand there was political progress (breaking the power of the old feudal order), on the other this political progress fell a long way short of *older* notions of political democracy associated with ancient Greece. Yet at the same time, the free market and the market in news and information, as well other genres (such as fiction) opened up a cultural space (the public sphere) in which social forces *outside* the dominant property-owning classes (the bourgeoisie as they came to be known) could press for a fuller democratic settlement. However, that extension of democracy took place only within the sphere of the political realm, a political realm that had undergone no fundamental change in its relationship to the property-owning classes ('property' here refers to ownership of the means of social production, namely land, industry and finances). Hence the extension of the franchise opens up a structural and long-term problem for political communication, as Hobsbawm shows us. And further, the division between political power and economic power that is unique to capitalism as a socio-economic system lays the foundations for the kind of political alienation one sees all around us today. As Ellen Meiksins Wood argues, the late nineteenth century extension of the electoral franchise actually begins to alter the meaning of the word 'democracy' away from its radical roots in the democracy of ancient Greece. Now democracy comes to mean electing a specialised class of political 'doers' who act on behalf of the majority, while at the same time, much of the meaning of 'freedom' associated with democracy gets transferred over to the market (Wood 1996: 234). To be 'free' comes to mean the ability to buy and sell in the capitalist market place, to have rights to private property and to dispose of it as you will. The nature of the market however was undergoing change at the same time that the extension of the electoral franchise was beginning to be fought for and won. These changes made the capitalist market ever less convergent with the idealisation of it as a realm of freedom. And it fundamentally altered again the dynamics between the economy, democracy and the media. This strand of the narrative can be seen best in relation to the media itself. An emergent working class media which began to give expression to a growing network of working-class public life (trade unions, societies, cultural organisations) was crushed by the market logic of rising production costs and advertiser based funding for the media (Curran and Seaton 2003: 29). Such trends raised barriers of entry to the market because start-up costs were high and tied the press into the needs of their major funders, the advertisers, whose main goal is to sell the products and the ideals of capitalism. In other words, the capitalist market place was becoming a very unequal place, where capital was growing bigger, more concentrated and more interconnected than ever before.

The fate of the working class media reflects the growing mid-nineteenth century trend towards the monopolisation of the media and its 'refeudalisation' around a narrow constituency of intersecting interests dominating economic and state life. Habermas argues that in the early history of the press, when its mere existence and survival was problematic and threatened, the press necessarily engaged 'in the struggle over the range of freedoms to be granted to public opinion' but once its existence was guaranteed and presence normalised, the press became increasingly orientated towards 'the profit opportunities for a commercial business' (Habermas 1996: 184). Where these conflicted with qualitative improvements to the media dimension of the public sphere, commercial interests were ineluctably prioritised, a trend charted by early twentieth-century sociologists and press critics such as Ferdinand Tönnies and Karl Krauss (Berry and Theobald 2006). The same story can be told again and again, and perhaps will be played out again in some form or another in relation to new media platforms.

The media's myth of itself as the 'fourth estate' – checking and scrutinising the activities of the state – fails to acknowledge the substantial structural linkage of interests which subsequently developed between the state and the corporate private media after the latter's initial 'heroic' phase (Petley 2009). One statistic should suffice to illustrate the corruption of the public sphere by the 'free' media in the US and its nexus of interests with the capitalist state. Polls show that approximately 80 per cent of Americans believed that Saddam Hussein was involved in the attack on the World Trade Center towers and other American targets on September 11th, 2001. Such a belief, against all the available evidence, clearly functioned in the interests of the Bush administration, which was seeking reasons to validate its war on Iraq which it duly launched in March 2003. A properly functioning media would have provided the American public with the information and analytical resources with which they could have subjected the Bush government's clever conflation of 9/11 with Saddam Hussein to rational critique and hence, theoretically, restraint (see Rampton and Stauber 2003 for a full account of media collusion with the Bush administration's propaganda machine).

The UK media's performance in the run up to and during the Iraq war was hardly much better. David Edwards and David Cromwell who founded Media Lens, a website dedicated to critiquing and holding to account the press and broadcasting corporations, have charted in detail how journalists failed to subject the pro-war case and the consequences of war, to critical scrutiny. As an example, they cite Andrew Marr's extraordinary performance, as the BBC's chief political editor, on the day US and UK troops took control of Baghdad:

> Well, I think this does one thing – it draws a line under what, before the war, had been a period of...well, a faint air of pointlessness, almost, was hanging over Downing Street. There were all these slightly tawdry arguments and scandals. That is now history. Mr Blair is well aware that all his critics out there ... aren't going to thank him – because they're only human – for being right when they've been wrong. And he knows that there might be trouble ahead... But I think it is very, very important for him. It gives him a new freedom and a new self-confidence. He confronted many critics... He said that they would be able to take Baghdad without a bloodbath, and that in the end the Iraqis would be celebrating. And on both those points he has been proved conclusively right. And it would be entirely ungracious, even for his critics, not to acknowledge that tonight he stands as a larger man and a stronger prime minister as a result. (Edwards and Cromwell 2006: 52–3)

The historic decline in the foundations for a critical media was an important check to the potential threat to the status quo that the gradual enlargement of the popular franchise might have posed to Western capitalism. If the public sphere laid the philosophical foundations for the popular franchise, the media component of the public sphere in its later monopolistic phase played a not inconsiderable role in ensuring that the franchise did not unduly threaten the established interests of private property or make the state more accountable to ordinary people rather than corporate interests.

Neoliberalism and New Labour

The extension of the electoral franchise combined with the special historical circumstances during the worldwide recession of the 1930s and the Second World War (1939–45) meant the establishment of a social democratic capitalism. This model of capitalism provided social protection and benefits for ordinary people and insisted on certain minimal obligations on the part of capitalism towards society. However, during the 1980s under the presidency of Ronald Reagan in the US and the Prime Minister Margaret Thatcher in the UK, a new model of capitalism consolidated itself (called neoliberalism) and along with it, a new model of policies intended to be the new norm, the new commonsense. This is the context for the *contemporary* crisis of representative democracy and the media crisis within that broader bourgeois public sphere which it helps construct.

Neoliberalism has been underpinned by and has facilitated the globalisation of capital, which has meant greater capital mobility, greater national receptiveness to penetration by international capital, new and speedier communication systems and the cheap transport costs necessary to ship commodities around the world. This has essentially brought enormous volumes of cheap labour from the developing world onto the world market, thus providing capital with the leverage to substantially dismantle the social gains won in the West during the social democratic compromise of yesteryear. Yet here we find a fundamental point of disjuncture between political and media elites who constantly tell their voters, readers and viewers that the welfare state is unaffordable, and large swathes of the population in the advanced capitalist countries, including the UK, for whom the prospect of competing with the labour costs and working conditions of Chinese workers is understandably unattractive. For although much of the media has conducted a long-term campaign to erode attachments to socially provided entitlements, and has sought to present the new neoliberal model as both a growing fact (i.e. inevitable) and a normative value (i.e. a good thing), there is a widespread passive and active refusal to agree that the commodification of public services is either necessary or desirable.

Across Europe the political class has been reconstituted around the neoliberal model – a fundamental consensus exists within and across all the mainstream parties that low taxation, low growth, relatively low public spending (as a proportion of GDP), privatisation, deregulation, the erosion of trade union power and liberalisation of trade constitute the cornerstones of public policy. But this consensus in the 'commanding heights' of political life within nations and amongst international bodies such as the EU, the North American Free Trade Agreement (NAFTA), the World Trade Organization, the International Monetary Fund, and so on, is very far from being matched by people on the ground. This has been particularly true in France, where the European social model has been defended more resolutely by progressive forces, than in the UK. The French populace has been in the forefront of defending the European social model, rejecting in a referendum a new European constitution which the entire French establishment was advocating, but which many others argued enshrined the Anglo-Saxon model of unfettered capitalism. This episode reveals how the contradiction between the international economic organisation of capitalism and its political legitimation grounded in nation states is now becoming profoundly destabilising (Amin 1997: 22).

The integration of political elites within nation states into a broad global agenda arguably sounds the death knell of the kind of national autonomy that once provided the foundations for the reform of capitalism. The nation state's legitimacy is supposed to reside in it being a reflection – either through democratic electoral politics or even (more mystically) in dictatorships – of the 'people', the conjoining of nation *to* state. It is precisely the legitimacy of the state and the unity of the nation-state relationship which globalisation throws into crisis. Gary Teeple calls globalisation the 'Second Bourgeois Revolution' (Teeple 2000). The first laid the foundations for the nation state, the second will produce a dialectical *aufhebung* (transcendence and incorporation) of the nation state into the order of global capital. The first revolution constructed a unified *national* market within a physical territory, complete with standardised laws, regulations, language, culture, political systems and the steady erasure of pre-capitalist social relations that were impediments to capital accumulation. The second revolution seeks to construct a unified *international* market, complete with substantial standardisation of laws and regulations, the international meshing of political systems into international para-state structures and the steady erasure or hollowing out of social relations associated with the social democratic past. As the circuits of capital move from the national to the global, the capacity of citizens to influence capital via national political institutions is substantially reduced. This is not to say that the nation state is in decline; indeed, it has played a crucial role in building the architecture of global capitalism. What has declined is its willingness and therefore capacity to provide checks and balances to the power of capital. Indeed politicians openly declare that their aim is to limit 'the influence of politics over the economic' (ibid.: 159). Globalisation has thus in many ways reaffirmed the traditional Marxist critique of bourgeois democracy and made it more relevant than ever. The merging of the state apparatus with the corporate monopolies, the bureaucratisation and militarisation of the state apparatus, the deterioration of the role of national political institutions (hardly compensated by the growth of regional political institutions which very largely internalise what is deemed 'realistic' politics set at a national and international level) and the extension of the powers of the executive bodies of the state, including the office of the Prime Minister which has come to resemble the Chief Executive Officer of a large corporate firm (Mishin 1975), have all grown apace during the neoliberal era. The UK has been particularly susceptible within Europe for neoliberal transformation because of its regressive electoral system, centralised state, dominant Executive and no official/written constitution (Leys 2001: 30–3). These structural political arrangements that made the UK vulnerable to the neoliberal virus are also underpinned by historical and cultural forces. England, the dominant nation within the UK, was of course the birthplace of those liberal economic practices and ideas that neoliberalism has revivified. The late twentieth-century return of classical liberalism in the mutated form of neoliberalism was thus nourished by the deep reserves of individualism that run through British, and especially English, culture, and which Margaret Thatcher tapped into so successfully. At the political level, this recrudescence of liberalism can be detected in the stated aim of New Labour's leader Tony Blair during the 1990s, to 'heal' the great split between liberalism and socialism that occurred with the birth of the Labour Party in the early years of the twentieth century. In other words, Blair's aim was that the distinctive set of interests and policies of the Labour Party, that were to some extent rooted in the organised working class, be folded back into the liberal tradition from which it had liberated itself, a tradition which has always privileged the freedom of the market and private property over the freedom of the workers (hence the traditional hostility of liberals towards trade unions).

This project to annul the break with liberalism that had actually founded the political party that Tony Blair was leading manifested itself for a brief time as the so-called Third Way politics. The Third Way was supposed to reconfigure Thatcherite policies and social democracy into a new synthesis, where economic efficiency and social justice could be reconciled (Giddens 1998) and there would be no need for the vocabulary and policies associated with socialism, in its Labour Party or any other variant. In fact the Third Way now looks, as a number of its critics identified fairly early on, as political cover to continue the neoliberal project. Under Tony Blair, New Labour developed the Thatcherite agenda of advancing the 'marketisation' of society, but in a manner that brilliantly disguises the process. Instead of a high profile and increasingly unpopular programme of killing the public services with hundreds of cuts in public spending, they were to be colonised by market forces. New Labour radically implemented a device brought in but rarely used by the previous Conservative government. Public Private Partnerships (PPPs) (a generic term for involving the private sector in some aspect of the delivery of public services) and especially Private Finance Initiatives (PFIs) (to build public sector infrastructure such as buildings) have funnelled large amounts of tax payer's money into very deep private pockets. Private Finance Initiatives, which essentially mean paying back a private company over, typically, a 30-year period, costs more for the tax payer in the long term but does not show up in the government accounts as public spending because the cost of the initial outlay is met by the private company. The contracting out of public services via PPPs and PFIs results in the penetration of public services by market forces and market logics, with predictable results for public sector workers and users. Evidence of the Third Way as a project re-branding the neoliberal agenda can be found in the very language of New Labour. Norman Fairclough's analysis of New Labour discourse finds that the multinationals rarely figure as *agents* in New Labour's view of the global economy. They are instead rather like 'ghosts in the machine', while the machine itself, the global economy, is presented as something that operates (with the use of metaphors) like a force of nature. The grammatical device that underscores this, argues Fairclough, is nominalisation – the use of nouns, rather than verbs. Nominalisation backgrounds the processes and the social agents that cause change and foregrounds their effects (Fairclough 2000: 26). For example, words like 'reform' (as in 'public sector reform') and 'modernisation' are usually used by New Labour as nouns. As such the sense of the government's own agency in public sector 'reform' and 'modernisation' is weakened. Instead these nouns background the processes of privatisation, deregulation and market competition, which underpin New Labour policies, presenting 'reform' and 'modernisation' as self-evidently good 'things', rather than activities (verbs) designed with particular outcomes that serve particular interests. It is the subtle difference of emphasis between saying 'we are going to reform the public sector' and 'we are committed to reform in the public sector'. In the latter case, 'reform' does not appear to have a causal agent, it is simply there or happening of its own accord. This is a *fait accompli* that has damaging consequences for public debate.

The refusal to engage with the market as a socially constructed institution that is alterable means that New Labour's political discourse is peculiarly 'closed'. As Fairclough notes:

> Although New Labour constantly initiates 'great debates' and calls for debates and discussion around its policy initiatives (e.g. welfare reform), it seems in broad terms that it sets out to achieve consent not through political dialogue but through managerial methods of promotion and forms of consultation of public opinion (e.g. focus groups) which it can control. The Government tends to act like a corporation treating the public as its consumers rather than as citizens. (Ibid.: 12)

Not long after Blair presidentially and single-handedly dragged Britain into supporting George Bush's 2003 war in Iraq, New Labour launched a series of 'national conversation' roadshows. Having ignored the majority view of the British and world public vis-à-vis the folly of that war, New Labour now wanted to show that it was 'listening'. The *Guardian* journalist Madeline Bunting joined one of these discussion seminars but found a 'clash of languages' between Blair himself, who was present and the vernacular language of the public:

> Instinctively, Blair uses the language of business and the market; it's not just a chance metaphor, it's how he sees the world; he has deeply absorbed the market ethos and believes that this is the language the public understands. He talked of the NHS's customers; he stuck firmly to his mantra of the benefit of choice in health and education to leverage up standards and of the private sector to innovate. (Bunting 2003: 26)

In contrast, the people in her group did not feel the market was the best way to make health care responsive to people, while another topic, education, was discussed in ways that sought to uncouple it from being mere preparation for the labour market. Indeed the whole exercise begged larger questions about the nature of representation and power:

> ... this was where the disconnect hit home – the audience had no power, it wanted power and wondered whether its deliberations would work their way into the manifesto. But beyond that, there was nowhere to take its pent-up energies, no sense of how this vibrant discussion could be translated into a political process. (Ibid.)

Critiquing the media or blaming the media

The argument thus far sketches out the historical, institutional and structural reasons underpinning the disconnection that exists in the UK from politics. We have seen that neoliberalism exacerbates what is already problematic in the structure of capitalism, namely the exclusion of economic power from political control and accountability. The social democratic 'interlude' provided some compensation for this with welfare entitlements, and some minimal state guaranteed obligations on the private sector towards society. The dismantling of this compromise arrangement and the simultaneous growth in the power of private property over politics and public life leads to a growing disillusionment in established political forms and channels. The new neoliberal state increasingly sees its populace as consumers rather than citizens, and so neither needs nor desires much in the way of popular involvement in political life.

Yet there are those who think that widespread disengagement with politics and hostility towards politicians by the public is largely the fault of the media. Such a view would seem to indicate that the media are positioned largely in confrontation with political elites. A common complaint amongst political elites across Europe is that the media are to blame for the poor esteem that politicians are held in and that the media are responsible for the public perception that politics is failing them (Nieminen 2007).

When Tony Blair stepped down as Prime Minister in 2007, he described the news media as a 'feral beast' (Blair 2007: 479) which obviously suggests a somewhat poisoned relationship between the media and the political class. This would seem to contradict some of the indications we have already given that the corporate media have long since cast off whatever role they once had to hold the powerful to account. Crucially, what we have to understand here is that the conflictual relations that can and do exist between the media and the political class do not rule out a deeper underlying consensus on fundamental policy issues and a shared conception as to what constitutes the 'natural' social order.

A good example of this coexistence of consensus and conflict is the episode involving the BBC and Tony Blair's government over the Radio 4 report by Andrew Gilligan in May 2003. The report suggested that the intelligence dossier on the weapons of mass destruction supposedly held by Iraq prior to the invasion of that country in February that year had been spiced up after political pressure exerted by government officials (specifically, it was alleged later, the Prime Minister's chief communications strategist, Alistair Campbell). Gilligan's report was based on a source inside the Ministry of Defence, who was later exposed as Dr David Kelly. The episode caused a major conflict between the government and the broadcaster, but it ought to be noted that the terms by which the broadcaster came into conflict with government were very much within an establishment consensus. First, the basis of the report was a secret source inside the Ministry of Defence, rather than any of the numerous voices publicly calling the basis of the war into question in terms of a broader argument concerning Western imperialism, resource wars and geopolitical struggles in the Middle East. Such voices had little profile within the mainstream media, and certainly very little within BBC reports about the war, before, during and after it. Gilligan's report and the furious response to it by the government assumes a belief in the separation of powers in the first place, so that it becomes 'improper' and 'scandalous' that political considerations might enter into 'intelligence matters'. Yet only a child or extremely naïve person could seriously believe that politics and intelligence matters are not integrally connected most of the time. The dossier on Iraq's weapons of mass destruction was called into question as a 'fix' by many people almost immediately upon its publication in September 2002, but such people were largely outside the charmed circle of the mainstream media, their sources and commentators, who were falling over themselves to be convinced of the case for war (with the honourable exception of the *Daily Mirror*).

The US example

Media critiques that blame the media for political disenchantment appear to have emerged first in the US, the spiritual homeland of neoliberal capitalism, the place where the media are the most corporate and the least regulated, and where there is only a tiny and marginalised public service broadcasting system. Here there have long been concerns about the so-called dumbing down of the political conversation. However, while there are legitimate concerns about news media performance, there is a clear political divide separating a critique of the news media's coverage of politics (in the broadest sense of the term) from blaming the media, a tradition that tends to isolate the media as the main issue, broadly absolve the political class and structures of any substantive problems, and resolutely refuse to connect political disengagement with neoliberal capitalism.

The implication is that if only the media would get its house in order, the public would come flocking back to mainstream politics. So it is hardly a surprise to find the neoconservative ideologues during the Ronald Reagan years of the 1980s blaming television for 'eroding respect for authority by exposing political scandals (as well as business corruption and failures) while fostering cynicism, distrust, and disrespect for the system as a whole' (Kellner 1990: 4–5). What happened, though, in the 1990s is that this habit of blaming the media (coming from sources with obvious political connections to the Reagan administration) crossed over into and mingled with longer standing academic critiques of the media. These critiques identify how impoverished the discussion of US politics is by the media generally and network television news programmes specifically. The critique focuses on the way television frames politics in terms of 'strategy'. This is to say that the news focuses on who is winning and who is losing (the 'horse race' analogy), it deploys a language of games, wars and competition, it obsesses with the 'performance' of political actors and measures the efficacy of any political position in terms of polling data. The cumulative effect of this is that substantive policy issues that would help voters make rational and informed choices are marginalised. Instead, there is a focus on individual performance during election campaigns, and outside those periods, a focus on the career trajectories of politicians as well as on scandals when they are found to have done something wrong or been found to be incompetent. Lichter and Smith found that 'policy issues were described only intermittently and superficially' (1996: 19); Lichter later noted how heavily mediated political coverage is by the broadcasters themselves at the expense of political discourse coming from the actual politicians (Lichter 2001: 14); Kerbel, Apee and Ross argue that 'the horse race, manoeuvring, strategic posturing, and personal hostility that are the hallmarks of network political news' is as prevalent on American public broadcasting television as it is on the commercial channels (2000: 9). Valentino, Beckmann and Buhr argue that strategy-based interpretations of the political process do weaken intentions amongst voters to participate and undermine confidence in political institutions, but that those with weaker ties to parties and less education are more adversely affected by such media frames than are party-political people and the highly educated (2001: 363). Recurrent findings of this sort build a compelling case against television. But there are a number of weaknesses in much of this sort of liberal analysis which are then amplified by an explicit 'blame the media' agenda. What is remarkably absent is any analysis of *why* there are these tendencies towards reducing politics to a spectacle of winners and losers, and stressing individual performance at the expense of substantive policy issues. Nor is there much concern with the wider ideological consequences of this kind of coverage. As Shanto Iyengar argues

> television's unswerving focus on specific episodes, individual perpetrators, victims, or other actors at the expense of more general, thematic information, inhibits the attribution of political responsibility to societal factors ... (Iyengar 1991: 5)

Thus while the media's individualisation of any problem or issue may be extremely uncomfortable for individual politicians caught in a media 'feeding frenzy' over some error or transgression, the broader social structures and practices which lie behind the specific episode remain protected by a screen of media silence. Just as the 'social' is missing from much media coverage, so it is missing from much liberal and conservative commentary *on* the media. For example, there is little analysis of the institutional and economic structures of the media or any elaboration of a theoretical framework that could make sense of these structures. Missing too is any account of the institutional relations *between* the media and politicians. Instead the relationship is conceived in only one direction: what the media do to political discourse. The issue of what politicians do to political discourse by, for example, 'negative' campaigning, spin, and the presentation of policies as essentially issues of management and 'delivery' is downplayed or missing entirely. All this means that the entirely valid criticisms of media performance can quite easily tilt into a position in which the media come to be blamed for turning people off politics. Thus Valentino, Beckmann and Buhr argue that issue orientated coverage would reveal 'candidate platforms as sincere attempts to identify and solve real problems' (2001: 351). We can see here how an insufficient critical foundation for media critique can quickly collapse analysis into an apology for the political elites. Political actors may or may not be sincere, but that is hardly the issue. Interestingly, Tony Blair constantly tried to ground his policies, especially on Iraq, around his 'sincerity', which rather missed the broader point as to whether whose and what interests would be served by pursing certain policy goals.

The most sustained academic attempt in the US to frame the issue of political alienation in terms of the media spewing out a corrosive cynicism comes from Cappella and Jamieson. However, their book, *Spiral of Cynicism: The Press and the Public Good* serves only to demonstrate the analytical impoverishment of their liberal philosophy. Without asking whether the category of 'cynicism' adequately accounts for wider public disenchantment from political processes, they posit instead, against a cynical media, a world of public officials doing the best they can and motivated in the main by the ethics of public duty. The problem with this Panglossian account is that it stays at the level of individual motivations and never offers a structural and institutional account of the context in which public officials operate, including the context in which values, ideas and belief systems become commonsensical and unquestioned foundations for action. This methodological individualism leads Cappella and Jamieson to make a number of very questionable judgements when it comes to expounding the moral virtues of public officials. For example, they argue that President Bill Clinton's support of the North American Free Trade Agreement (NAFTA), despite its unpopularity amongst significant public constituencies such as trade unionists, showed political integrity by rising above 'electoral self interest'. They also applaud bi-partisan consensus building between President Clinton and a Republican Congress, which agreed 'that deep cuts in Medicare are required to balance the budget' (1997: 24). Cappella and Jamieson seem entirely unable to see patterns in political behaviour and to perceive the social interests with which Clinton has aligned himself in both cases. That NAFTA was a charter for the free flow of capital across national borders (it actually sparked an armed insurrection by the Zapatistas in southern Mexico) and threatened jobs, wages and conditions within the US seems completely to have escaped the authors' notice. NAFTA is in fact a good example of how political disenchantment is fostered when the constituencies who vote for politicians see them captured by the powerful interests and lobbying organisations of big capital. Likewise, the consensus between a Democratic President and a Republican Congress over the need to cut the Federal budget to Medicare, again suggests an alliance between the state and the private health sector against the 47 million or so Americans without health care insurance.

Blaming the media in the UK

The idea that the media are responsible for political disengagement in the UK is a common one. The argument usually takes one of two forms. Either the media are seen as so dominated by entertainment and 'soft' news that audiences simply do not know or care enough about politics to want, or indeed be able, to engage in the political process. Or the media stand accused of taking such a negative approach to politics that audiences are comprehensively alienated from the whole business. As the late Robin Cook MP put it to the Power Commission: 'If you always serve up to the public the sense of the political process as one of failure, then it's hardly surprising that you lose confidence in your democratic process' (Power Report 2006: 67).

It would be hard to deny that the media landscape has undergone significant changes since the 1980s. As Bob Franklin argues

> Journalism's editorial priorities have changed. Entertainment has superseded the provision of information; human interest has supplanted the public interest; measured judgement has succumbed to sensationalism; the trivial has triumphed over the weighty; the

> intimate relationships of celebrities from soap operas, the world of sport or the royal family are judged more 'newsworthy' than the reporting of significant issues and events of international consequence. Traditional news values have been undermined by new values; 'infotainment' is rampant...News media have increasingly become part of the entertainment industry instead of providing a forum for informed debate of key issues of public concern. Journalists are more concerned to report stories which interest the public than stories which are in the public interest. (Franklin 1997: 4)

It would, however, be equally difficult to claim that 'hard' political news has simply ceased to exist in both written and broadcast form; indeed, thanks to the growth of 24-hour news channels and of Internet news sites there is now, in fact, more of such news than ever in circulation. But could it be that, as Robin Cook suggested, there are qualities even in 'serious' political reporting which contribute to public disengagement from the democratic process?

The main proponent of this thesis in the UK is John Lloyd, a contributing editor to the *Financial Times*, a former editor of the *New Statesman* and Director of Journalism at the Reuters Institute. Writing in *Prospect* in 2002, Lloyd argued that one of the most potentially damaging aspects of contemporary society is the belief in media circles that

> there can be no grounds for trust in political institutions and representatives – not as a result of specific malfeasance, but as an *a priori* assumption. The famous advice of the then *Sunday Times* editor Harry Evans to his journalists: 'Always ask yourself when interviewing a politician – why is that bastard lying to me?' has passed from radical fearlessness to a commercial strategy with big implications for the health of our public life. (Lloyd 2002: 52)

Two years later, in his book *What the Media Are Doing to Our Politics*, Lloyd expanded on this thesis, arguing that the media now routinely treat politics as a 'degraded profession' (2004: 8) and as 'sordid, drained of content and increasingly unpopular' (ibid.: 24), whilst working on the 'cynical assumption that politicians are born liars and rogues' (ibid.: 10–11) and members of a 'debased class' (ibid.: 14). In Lloyd's view, the media have collectively decided that 'politics is a dirty game, played by devious people who tell an essentially false narrative about the world and thus deceive the British people' (ibid.: 20). The result is that British political journalism exhibits

> a pervasive contempt for the governing classes, especially for politicians; a definition of all official or corporate public relations briefings as 'spin'; a concentration on process at the expense – often to the complete obliteration – of policy and outcomes; a privileging of conflict and complaint and a dramatisation of ordinary conflict, such as within government, as crises, irrespective of any tendency to contain the conflict within compromise. This produces the opposite of what a journalism devoted to public understanding would wish to produce. (Ibid.: 195)

Lloyd thus concludes that the way in which the media report politics actually poses a threat to democratic institutions by encouraging at best apathy and at worst antipathy towards them in the population at large.

This is a considerable charge sheet, but also a problematic one. First, Lloyd's failure to make distinctions both within press journalism and between press and broadcast journalism frequently leads him into excoriating 'the media' with far too broad a brush. (Paradoxically, this is normally a charge levelled by journalists at media studies academics.) In particular he fails to make the crucial distinction between political coverage which appears to be motivated solely by a powerful animus towards the current Labour government by stridently Conservative papers such as the *Mail*, *Telegraph* and *Express*, coverage which appears to rest on the premise that Labour has had the temerity to usurp the 'natural' party of government; political coverage by the rump liberal press (namely the *Guardian*, *Independent* and *Observer*), which are, at least in parts, dismayed and alienated by New Labour's abandonment of liberal social values and its embrace of neoliberal dogma in the economic sphere; and, finally, more general political coverage which is merely properly sceptical and enquiring in its approach to the subject. To argue that political journalism as a whole, produced as it is from different ideological and political perspectives and from within different kinds of institutions, treats politics in a negative, destructive and alienating fashion is a considerable over-generalisation, and ignores significant differences within and between the British media.

The second major problem with this approach is that it puts all the blame for what it regards as the debased state of political reporting, and for the concomitant disengagement of sections of the British public from the political process, on the media themselves. The idea that the state of the British political system is to a considerable extent responsible for the way in which its workings are reported by the media, as well as for public apathy or antipathy towards that system, is barely considered.

Sentimental and rhetorical flourishes about Westminster being the 'mother of parliaments' notwithstanding, Britain's parliamentary system has long been open to the charge that it is archaic, elitist, unaccountable, and out of touch with the people whom it is supposed to represent. Recent events such as the expenses scandal of 2009 have simply served to give these criticisms wider currency and greater legitimacy. It can also be persuasively argued that modern communication trends, far from bringing politicians and the political process closer to the population, have actually distanced them still further.

In the 1980s, the Thatcher government initiated a radical shift in the way in which policies were presented to the public. Drawing on the skills of former journalists such as Bernard Ingham, PR specialists like Tim Bell, and the former television producer Gordon Reece, a formidable machine was assembled to regulate the flow of political information to the media. Politicians and policies were 'packaged' for media presentation and public consumption to a hitherto unprecedented degree. As Lord Young, then minister at the Department of Trade and Industry put it: 'the government's policies are like cornflakes, if they are not marketed they will not sell'. Significantly, this *aperçu* saw the light of day in the magazine *PR Week* (16 March 1988).

This process, which came to be known as 'spin', was then carried on by New Labour, first in opposition and then in power. Of course, it could convincingly be argued that New Labour was in dire need of an effective spin doctor such as Alistair Campbell, given that most of the press had been absolutely besotted with Thatcher and bitterly hostile to any who dared to oppose her – including members of her own party. However, far from taking on the hostile press, the New Labour spin machine did its utmost to placate it as part of the party's overall strategy of repositioning itself to appeal to 'Middle England'. Furthermore, as part of that strategy involved adopting the Thatcherite stance of regarding people as consumers first and citizens second, or, alternatively, of reinterpreting the notion of citizenship in thoroughly consumerist terms, it was perfectly happy to continue, and indeed to intensify, the process of marketing its policies like cornflakes. The result is that politics has become more packaged than ever, with consequences which are extremely damaging to democracy. As Bob Franklin argues

> Packaging politics...impoverishes political debate by oversimplifying trivializing political communications. Negative attacks on opponents' policies have come to be preferred by many politicians, as well as judged to be more effective, to the positive elaboration of their own case. Packaging politics places a premium on personalities and presidentialism. Persistent television portrayals of a select clique of 'telegenic' politicians, mouthing pre-rehearsed slogans and automaton soundbites, has supplanted the rational and sustained advocacy of politics. (Franklin 2004: 12)

Of course, in this process, politicians and the media are to some extent complicit. Politicians need the media in order to get their messages across to the public, and journalists need stories, not simply in order to keep the public informed but also to boost their organisation's (and their own) reputation as a reliable, well-informed, up-to-the-minute news source which is able draw on useful insider contacts. Thus, as Franklin points out: 'the relationship between politicians and journalists is best viewed as an exchange relationship in which insider political information and opinion is traded for coverage in news media...Each set of actors is crucially dependent on the other and the packaging of politics reflects, and in turn, nurtures this mutual reliance' (ibid.: 15). Politicians have to adapt their communications strategies to changes in journalistic practice, but, equally, journalists have to adapt their practice to changes in politicians' communications strategies. It is thus far too simplistic to blame the media for the manner in which politics is reported, and thence to ascribe blame for political disengagement to the press and television.

As the Power Report suggests, the problem is much more complex than that. Thus, it argues, on the one hand:

> Recent governments, all too aware of the power of the press, have tried to bend it to their own purposes and in doing so have shown little respect for what journalists ought to be doing in an open democracy. The media is treated as yet another part of the mechanics of government to be bent to governmental will. Not surprisingly this leads to mutual disrespect. (Power Report 2006: 68)

On the other hand, partly as a result of this disrespect, but also due to factors deeply engrained within the press culture itself, the press tends towards a simplistic, Manichean view of political life which leads it to be 'reductionist about quite complex issues and sensationalist about failure or flaws. Bearing the head of a politician or celebrity on a pole is the ultimate victory in the ratings wars and this feeds into the atmosphere of disdain for those in public life' (ibid.). Of course, the broadcasters are supposed to be impartial in political matters, but, given the newspapers' powerful agenda-setting function, it is well-nigh impossible for them to remain entirely unaffected by what appears in the press.

However, many of those who submitted evidence to the report felt that politicians were not sufficiently accountable within the political system itself and thus it was right that the media held them firmly to account. This attitude clearly reflected a wider scepticism about politics and politicians, and the report thus concluded that negative media coverage of politics is a *symptom* as opposed to a significant *cause* of political disengagement and alienation: 'It seems to us that while there clearly is a problem with the media the answer is where to start and, in light of the evidence, we believe that if we get the political system right this will change the atmosphere and culture and the press will follow' (ibid.: 69). This is a judgement with which, in the light of the ideas already expressed in this chapter, it is difficult not to concur.

Conclusion

The contemporary crisis of political representation has deep structural roots. Although the contemporary moment is conditioned by immediate changes in the nature of a globalising capitalism, a historical perspective reminds us that the relations between democracy and capitalism have always been historically variable and, with the division of labour between political representation and political power on the one hand and economic power on the other, deeply problematic. The historical trajectory of the media, from organs of public expression against the undemocratic and unaccountable power of the state, to corporate entities prioritising commercial considerations and identifying their interests with the capitalist state, means that the media's relationship to the public and the political elites is far less productive of democratic debate and information than the media's own idealisation of itself would suggest. Political discourse itself has been corrupted by the shrinking ambition and capacity of political elites to offer any choices outside further embrace of and penetration by market forces to the detriment of hard-won social entitlements. Political consensus at the top of society has eroded real effective choices for voters. The absence of ideological divisions within the political class must go a long way to explaining the paucity of interrogation by the media of policy choices and policy-making, especially when the media are so reluctant to go outside the establishment for interpretive comment and analysis. The tendency towards personalisation, individualisation, immediacy, dramatisation, simplification, titillation and conventional thought (Chibnall 1977: 23–39) when it comes to political coverage must in part be understood as an attempt by the media to fill the vacuum created by this ideological unity. Otherwise where is the drama of political debate? However, these categories of interpretation are also readily available to the media because their own commercial and ideological imperatives push them in this direction and away from proper social and systemic analysis. The contradictory effect which this has, however, is to reinforce the sense of our political representatives as irrelevant, self-absorbed and self-interested, without fostering amongst the pubic at large the intellectual capacities to reform the political process, much less find alternatives to it.

2
Young People, Politics and Television

We have seen that the capacity for representative democracy to be representative of social interests other than capital is on the decline. Therefore, the rationale for popular involvement in and representation through the political process is being eroded in the minds and actions of many social groups within Western democracies, including young people. After two world wars, the winning of the popular franchise laid the basis for the social democratic Keynesian welfare state in Europe and post-Roosevelt (and later after the black civil rights struggle) a similar if somewhat watered down social democratic model in the US. The establishment of the social democratic settlement in the UK after the Second World War had removed young people from exploitation in the labour market by providing universal state education and other welfare benefits. Phil Mizen argues that young people were one of the biggest beneficiaries of the post-war strategy of inclusion (Mizen 2004: 17). In underscoring the break with this strategy of inclusion and universal support inaugurated by the new neoliberal dispensation, Mizen is arguably a little uncritical of the social democratic model. The welfare state was still a compromise between capital and labour after all and the state still undertook to mould the latter to the needs of the former through its various apparatuses, such as the education system (Willis 1977). The neoliberal model however sees a shift from state provision and certain minimum guarantees of care and entitlement, to market provision – with all the substantial widening of inequalities that implies – and individual responsibility for satisfying needs and finding opportunities.

In the UK, income poverty, measured by households with less than half the national average income, more than doubled (from 6–12 per cent) in the latter half of the 1980s, as the consequences of Thatcherite policies began to take effect. Although they have since dropped to approximately

8 per cent by 2005, this is still 2 per cent higher than in the mid-1980s (OECD 2008). With the rising level of unemployment across Europe as capital becomes more footloose and fancy free, opportunities to work have been declining.

> In contrast to the two thirds of young people leaving school to directly enter work at the beginning of the 1970s, 25 years later only 8 per cent were finding work. By the time the Conservatives lost power in 1997, only one in five minimum age school leavers were entering the labour market directly and half of these were still destined to begin their working lives on government training or work experience schemes. (Mizen 2004: 55)

The school-to-work transition has thus become more protracted, more complicated, more risky and more diverse in trajectory and outcomes than was the case in the more relatively standardised, homogenous and secure model of the social democratic period. The crisis of working-class communities in terms of secure employment and good housing, combined with the difficulties of organising effective trade union protection within a panoply of restrictive laws and amid the new more fragmented service sector, along with family breakdowns and dispersals and geographical mobility, means that 'young people are less likely to have or experience vital cues for engagement' in traditional political processes and loyalties (Kimberlee 2002: 95).

Furlong and Cartmel describe this shift from a culture of solidarity to fragmentation and competition, as the 'individualisation of risk' (1997: 4). This is a process of social atomisation that breaks down both the solidarities essential for collective agency (especially through trade unions) and the *consciousness* of the structural and systemic determinants that produce – despite all the 'diversity' and heterogeneity of contemporary life – differential (especially class determined) outcomes in terms of life chances. These changes in the life experiences of a new generation ought to be mapped onto the long-standing 'life cycle' explanations of young people's disengagement with politics. In other words, it has long been true that people develop an interest in politics as they grow older, enter the job market on a more permanent basis, pay taxes, get married, have children, and so forth. Thus at one level disengagement from politics for young people is not some thing new.

Stradling's report for the Hansard Society in the late 1970s found in a survey of more than 4000 15–16-year-olds that their political knowledge was not sufficiently well formed to enable them to know how they were affected by politics or for them to have political agency themselves. What 'political knowledge...they do have is of a rather inert and voyeuristic kind' (Stradling 1977: 57). Another study drawing on fieldwork done in the 1980s with working-class young people found that 85 per cent of comments about politics were of the 'politics is boring type'. Again, respondents largely defined politics as party politics (Bhavnani 1991: 139). This disconnection from politics in the heyday of Thatcherite attacks on the working class may seem surprising. Yet in a society with a strong division of labour between a professionalised class of political 'doers' and the rest, and in a society where being young involves high degrees of socially constructed dependence on and conformity to institutionalised hierarchies of power (both in the home and in relation to education) one could hardly expect there to be a golden age of youth participation in politics.

Nevertheless, the new determinants shaping the post-social democratic generations do appear to be pushing disconnection from traditional political processes deeper and wider. At the UK election in 1992, the names of more than a million 18–24-year-olds were absent from the electoral register (Furlong and Cartmel 1997: 101). This was followed by a dramatic fall in participation in the national elections of 2001 where MORI estimated that only 39 per cent of 18–24-year-olds voted (Russell et al. 2002: 6). In the 2005 general election only 37 per cent of 18–24-year-olds voted (Power 2006: 34). Such trends brought much indignant commentary in the popular press, bemoaning the apathetic youth who were apparently too lazy and indifferent to participate in the democratic process (Wring, Henn and Weinstein 1999).

Within the mainstream media, such causal factors underpinning this disconnection such as the neoliberal consensus amongst the main parties and the emasculation of social democratic reform in the context of globalisation were largely off the agenda. Instead it was claimed that the *style* of political debate was too adversarial and that this turned the public off (a debate that conceals the extent to which the *substance* of politics is already *too* consensual), or that voters were a bit lazy and that it needed to be made easier for them to vote with new methods such as postal voting, or that the media themselves were encouraging cynicism and political apathy with the way they covered politics.

From the political class, one response to this has been the concept of 'active citizenship' which emerged in both the discourse of Conservative governments from the late 1980s, through to New Labour in the late 1990s, to try and rationalise and legitimate this new social atomisation and at the same time reconcile it with retaining *some* conception of society at all (Hall, Williamson and Coffey 2000). The 'active citizen' is one who accepts the personal responsibility and personal choice agenda of the neoliberal settlement. Here they are either active consumers or atomised workers selling themselves in the labour market, or undergoing retraining or educational programmes. Labour market employment and educational attainment have been key goals by which New Labour has sought to integrate marginalised young people into society. But the individualisation of risk means that policy frameworks cannot address the wider causal factors behind marginalisation:

> Whilst these [employment and education] are clearly critical with regard to young people's long-term prospects, much less attention is given to investigating young people's material and social circumstances, their incomes, living conditions, and the wider structural terrain of (dis)advantage which shapes their opportunities, choices and prospects. (Fahmy 2008)

In terms of media representations, individualisation of risk leads to the decomposition of our ability to attribute societal causes to various effects (such as crime). Individual behaviour and individual agency comes to be seen as the prime determinants behind crime, as if behaviour and motivation develops in a social vacuum. Marginalizing wider structures of inequality helps to reinforce citizen identifications with an expanding state apparatus around the theme of 'law and order' (as in the area of crime and increasingly, the criminalisation of political protest).

Despite such evasions, the sense of crisis in the capacity of representative democracy to represent a social base wider than big business has been evident in the growing number of quite popular books investigating the emasculation of democracy. Will Hutton's (1996) *The State We're In* suggested that the damaging domination of short-term profit motives that prevented companies investing in 'human capital' was peculiar to a Thatcherised UK. This was followed up by more radical critiques of the state, such as George Monbiot's (2001) *The Captive State* which charted the penetration of the British state by the forces of big business (rather than it simply being, as Hutton had it, a question of choosing a different, non-Thatcherite policy). Paul Foot's *The Vote: How It Was Won and How It Was Undermined* offered a historical account of how the democratic political franchise was won in Britain over several centuries and how it was gradually denuded of meaning culminating in the birth of New Labour. Commenting on the New Labour philosophy as articulated by two of its ideologues, Roger Liddle and Peter Manelson, Foot notes:

> The role of an elected Labour Government was no longer to intervene in the economy in the interests of the people who voted for it. Instead, its role was to stand aside from disciplining the rich and powerful, and reserve its indignation and discipline for the people at the bottom of the pile. The effect on the democratic process was devastating. (Foot 2005: 414)

Hilary Wainright's *Reclaim the State: Experiments in Popular Democracy* (2003) meanwhile investigated various experiments in popular democracy at local levels in both Brazil's Porto Allegre, famed for its participatory budget, and the UK. The widespread recognition of the depth of the political crisis may be measured by the fact that in the London borough of Harrow, not far from where research for this project was conducted, the local council, in conjunction with the Roundtree Trusts' Power Inquiry, have developed their own version of the participatory budget (Giddy 2005).

The new politics

What the mainstream debate about voter disconnection does not encompass is the openness of young people towards political engagement that is outside formal party politics. Beyond the superficial commentary within the media about an 'apathetic' youth, more considered research showed a different picture. Studies into young people's attitudes have found that for them

> Politics is often taken to mean 'party politics' or life in Westminster, but at the same time there appears to be a real interest in, and knowledge of, a wider set of political themes. There are signs of undoubted alienation but also a desire for much more 'information'. (Russell et al. 2002: 8)

Similar conclusions were reached by White et al. in their study, including the suggestions from young people that the media coverage of politics focuses on issues relevant to them (White, Bruce and Ritchie 2000). Tom Bentley and Kate Oakley's *The Real Deal, What Young People Really Think about Government, Politics and Social Exclusion*, consulted poor, unemployed and homeless young people and found similar results:

> These young people are alienated from mainstream politics and distrustful of politicians in general. They feel they are labeled and discriminated against by adult society. (Bentley and Oakley 1999: 8)

As with other studies, however, young people are not necessarily turned off from politics *per se*. Bentley and Oakley found evidence of interest in new 'forms of political engagement' (1999: 14) in the same sample based on 'single issue' campaigns and involving non-party based organisations. For these young people, the consensus amongst the main political parties meant that 'the voting choices they faced did not reflect the structure of their concerns or preferences' (ibid.: 119). In this respect young people are merely 'a barometer for wider social change and for the state of society as a whole' (ibid.: 13).

One of the most visible manifestations of a progressive politicised response to the crisis of mainstream politics as a vehicle for reform, let alone radical change, emerged in the late 1990s, with the anti-globalisation movement, the social movements, the anti-capitalist movements or globalisation 'from below'. Young people have been at the forefront of this non-party based politics of revolt and non-violent direct action. In the West this new form of political action has been composed largely of the educated and/or middle class (Norris 2002: 201) as well as a mixed category outside or in a peripheral relationship to the labour market, such as the unemployed, students and young people, housewives and the retired (Offe 1985: 831–2). The seeds of the anti-globalisation politics in Britain were sown by a disparate assortment of sometimes overlapping causes, activists and protests, many of which had the kind of articulation between youth culture and political protest that had perhaps been found to be only latent in the subcultures of the 1960s and 1970s (Hebdige 1979). In the 'pleasure/politics zone of party and protest' (McKay 1998: 45) or cultural production and political action is to be found the converging networks in opposition to The Criminal Justice and Public Order Act (1994), action around animal rights, anti-road protesters, free parties, Situationist style interventions into public space (Reclaim The Streets), Video-activism (Undercurrents) and then a growing internationalism that finds a key focus in the exploitation of the developing world. These organisationally fluid and informal networks took on new targets (supra national and corporate) and sought to open up vast swathes of social life to political contestation after the neoliberal attempt to reconstitute the economy, the market, property, commodification and consumption as uncontestable and non-political features of life (Offe 1985: 820).

Curiously a number of theorists in the 1980s and 1990s talked about the emergence of a 'post-materialist' politics (Inglehart 1997) or a 'life-politics' (Giddens 1991) to distinguish it from the apparently 'economistic' or 'bread and butter' issues associated with trade union based struggles. For Habermas the 'welfare-state pacification of class conflict' (1987: 348) means that political conflicts shift away from questions of economic ownership and resource distribution and over to issues thrown up by the colonisation of the norms and values of everyday life (the lifeworld) by the system worlds of economy and state. A new politics centred around 'how we live' and the morality and values of life emerges according to Habermas. Leaving aside the issue that such 'how we live' questions have always bubbled away within the 'old' politics of the labour movement, especially in the more radical, less economistic struggles, the erosion of the welfare state by neoliberalism and the penetration of the lifeworld by an increasingly rampant corporate capitalism has reintegrated many of the 'post-material' politics back into recognisably materialist concerns (Edwards 2004). What was seeded in the 1990s was a growing attack on the corporate culture of capitalism and a growing inability of traditional political channels to express this attack.

Naomi Klein's (2000) *No Logo* represented in this regard, a virtual biography of a young generation awakening from what Walter Benjamin called the phantasmagoria of consumerism. What was nascent in Benjamin's studies of the Paris shopping arcades in the nineteenth century (Benjamin 1999) had a quite specific historical restructuring and expansion towards the end of the twentieth century. Klein charts the shift of multinational corporations towards the production of *brands* and away from the production of goods, the latter now, in a globalised economy, contracted out to sweatshop labour in the newly industrialised or developing countries. Branding is more than advertising, although advertising spend is crucial to its existence. It is the imbuing of goods and associated corporations with 'culture', with a whole way of life, with meanings and values. It is '[m]arketing that thinks it is culture' (Klein 2000: 66) in order to turn 'lacklustre products into transcendent meaning machines' (2000: 68). This 'symbolic turn' or shift to the production of meanings in the advanced capitalist economies, and the omnipresence of branding as well as the centrality of both the youth market in itself and youth cultural images across all markets, led, in a welcome dialectical response, to the politicised interrogation of the discrepancy between the shimmering brands and the coercion, desperation and violence involved in the production of the goods to which brands attached their transcendental idealism. *No Logo,* which begins with a critique of commodity fetishism, ends with the struggle of workers to organise themselves and resist exploitation. Thus global capitalism has, despite itself, produced elements of a globalised consciousness. Amongst the big corporate casualties of the practical politics *No Logo* springs from and charts was the fast food giant, McDonald's, whose libel action in the UK courts against two anti-McDonald's campaigners amounted to the largest public relations own goal in corporate history. The pressure on McDonald's was further ratcheted up by the worldwide success of Morgan Spurlock's critique of the corporation's activities in the US in his documentary, *Supersize Me* (2004). In the UK, *McLibel* (Franny Armstrong 1997/2005) charted the libel action in the courts from the 1990s, but was subsequently updated and finally broadcast on BBC Four in 2006.

Young people and well-being in the UK

Yet although one can point to political engagement by many young people today, as in the high-profile involvement of school students in protests against the war in Iraq (ritually denounced by politicians, educationalists and the media) it is also evident that for many others, political disengagement from any form of politics *is* the norm. Here the class profile of the disengaged tends towards those who have the least opportunities in terms of education and employment, who live in poor housing conditions and are often subjected to racial discrimination. In such circumstances of pressing needs, the absence of media representation, of information and of media channels to articulate social problems and discover potential solutions combine to further frustrate the possibility of addressing social grievances. In the UK young people are managed and contained by an increasingly authoritarian state, underpinned by the legislation of successive governments policing youth's everyday space (ASBOs, Stop and Search, on-the spot fines) and a broader culture of suspicion and fear towards young people. Just as the neoliberal state becomes more disciplinarian in relation to workers and provides a legal framework for managing labour/employer relations that weigh massively towards the latter, so the state becomes increasingly authoritarian vis-à-vis youth.

Across numerous indicators life is becoming harder for young people in a country that has unravelled the social democratic consensus. A report by UNICEF in 2007 measured child well-being in affluent countries using data compiled under six categories. Under 'material well being' the report looked at data on relative income poverty, households without jobs and reported deprivation. Under health the report looked at mortality rates from infants through to 19-year-olds as well as vaccination programmes. Educational well-being looked at educational achievements at age 15, at post-compulsory education and the transition from education to employment. Under the category 'relationships', the report looked at family and peer relationships. Under the category 'risk' the report drew on data regarding health (food intake and physical activity), consuming harmful substances or engaging in potentially risky sexual activity, and experiences of violence or bullying. Under the final category of subjective well-being, the report compiled data on how young people themselves feel about their lives, their health and their school life. What is interesting about this study is that the two affluent countries that have most embraced neoliberal capitalism (in other words, to embrace a society where there is affluence for some and poverty for many) the US and the UK, came, respectively, second from last and bottom of the table when the marks across these indicators were averaged out (UNICEF 2007: 2). In 2009 a report by researchers at the University of York for the Child Poverty Action Group came to similar conclusions. Here the UK came 24th out of 29 European countries in a league table of child well-being. Only Romania, Bulgaria, Latvia, Lithuania and Malta did worse.

Television and cognitive development

It is noticeable how much of the already existing literature on audiences and their relations with the media focuses on children, as opposed to young people. But studies in developmental psychology and media effects clearly have a broader relevance. Deborah Roedder John, for example, offers a three-stage model of socialisation in relation to advertisements for children and teens (2002). The model moves from identification of characteristics (perceptual stage, ages 3–7), to an understanding of how these characteristics interrelate into a coherent intention (the analytical stage, ages 7–11), to a contextualisation of the text in broader patterns of behaviour, status and cultural meaning (the reflective stage, ages 11–16). The movement 'outward' is a typical feature in cognitive development studies. However, it is precisely this process of building up the 'bigger picture', or in more theoretical terms, constructing the totality of social and economic relations, which becomes progressively unfulfilled and uneven in no small part because of the media environment. The media environment plays a crucial part in mediating the individual's experience with events and processes and systems that are either not experientially present to the individual, or whose causal dynamics (as opposed to effects) are not immediately perceptible. The impoverished quality of that mediation constitutes the key issue. The media in other words fail to (adequately) mediate.

The continuities between the findings of contemporary studies of the media (and contemporary events the media mediate) and R. W. Connell's study, conducted in the late 1960s, are remarkable. Connell's ethnography of Australian children was both an example of developmental psychology and a sociology of politics. Although not specifically a study of television, his interviews with 119 children between the ages of 5 and 16 testified to the centrality of television in providing children with images and information about the world of politics. Connell's study was spurred by the mystifying political longevity of the then Australian Prime Minister, Sir Robert Menzies.

Connell also has a three-stage model of cognitive development. At the Intuitive stage, 'children lack a conception of political structure, not because they lack sources of information about it, but because they lack the cognitive equipment to represent it' (Connell 1971: 18). At this stage then (up to the age of 7) information provision is far in advance of cognitive capacities but provides important stimuli for the latter. In the second stage, termed Primitive realism, there is a growing sense of what constitutes political authority and power, although there are many confusions in figures, tasks, titles, rules, and so on (ibid.: 24). In the third stage, the construction of political order (around age 10–11 onwards), a recognisably embryonic adult conception of political order, begins to emerge (ibid.: 38). A picture of hierarchies, power, levels of government, parties and conflict over policy is gradually fleshed out. It is also here that the young person's thinking about the world begins to evolve into recognisably ideological configurations, at least about the key issue of the times, that is, Australian involvement in the Vietnam war (one might suggest that other modes of ideological thinking grounded more immediately in personal experience, such as gender relations, come a good deal earlier). Connell found that on the one hand the 'children were almost unanimous in their view of the war, and almost unanimous in their condemnation of it' (ibid.: 95). And yet, strikingly, those 'in favour of Australian involvement outnumbered those against by two to one' (ibid.: 96). Here we have the clear beginnings of consciousness beset by political contradictions (as opposed to mere confusion) that derive *not* from the incomplete process of cognitive development, but from social (and media) sources. Thus, while against the war at one level the children were 'for' Australian involvement because of a perceived 'threat' to their lives from the Vietnamese or from communism more generally (ibid.: 97). Connell found that television was a prime source of the children's knowledge about politics and his conclusions remain very relevant today. Television, he notes, encourages

> involvement of emotional reaction, not an involvement of action or potential action ... Television can show things to fear, things to be shocked by, things to amuse, things to like and things to hate, but it does not show the children things to do, forms of engagement. It simultaneously draws them in and holds them at a distance. (Ibid.: 129)

His more general conclusion is that

> when we consider the general tenor and the total content of the interviews, we must be impressed by the conventionality, the lack of *realized* freedom, in them ... the mass of the people is growing towards adult involvement without sign of the kind of practical imagination which would let its members generate plans of action to deal with the political problems they recognize. (Ibid.: 239–40)

Communication rights, or why turkey twizzlers are not ideal

Graham Murdock has argued that as part of the rights of media audiences, they should be presented with diverse experiences that are relevant or are made relevant to their lives; be able to see representatives of themselves within the media; be able to access information, especially in relation to 'the actions, motivations and strategies of significant social, political and economic actors'; be offered alternative frameworks of knowledge construction about the world and have rights to participate in the making of the media themselves (Murdock 1999: 11–12). However, the philosophical core of liberalism (whether of the neoconservative kind or the socially 'concerned' kind) makes it very hard to sustain an argument for 'communication rights' as Peter Wilkin has noted (2001: 81). For liberals it is the rational choices of individuals acting within the market place that provides goods to meet their demands and desires which ought to decide what is produced and consumed. Within this vision, to talk of 'rights' is to suggest that individuals do not necessarily have the means (whether material, informational, cultural or whatever) to identify which choices it might be good to have on offer. To relocate individuals within their structural contexts of determination 'decentres' the liberal model of the individual subject and undercuts the fantasy of atomised individuals pursuing their desires in the market place without significant effects on others or themselves.

In the Channel 4 television programme *Jamie's School Dinners*, celebrity chef Jamie Oliver struggles to overcome the attachment of schoolchildren to a diet of pizzas, chips and turkey twizzlers, while overcoming their resistance to even *trying* the healthier options he trains the dinner ladies to offer. Within the liberal model of individual agency, the schoolchildren have made rational choices about their food consumption and to speak of the right to have a healthier diet smacks of authoritarian interference in individual liberty. Erased from the liberal world view, however, are the complex ways in which the rational choices of the 'free' monad have already been pre-structured out of a range of social determinants, ranging from parental cultural capital, the rise of two-parent working families and the associated phenomenon of 'convenience' food, the contracting out of school dinners to private providers, the lack of training for dinner ladies and the barrage of advertising for cheap, high-salt, high-fat, processed and ready prepared food aimed at children.

The argument for 'communication rights' requires addressing those wider structures and collective failures that already circumscribe the individual's 'free' rational choice in the market place. The argument for such rights addresses the possibility that existing communication provision cannot be changed for the better simply by relying on the crude quantitative market signals returned to providers by consumers. Indeed it is not easy to see how providers, should they wish to do so, can interpret market signals on goods that by and large do not exist in the first place. The argument for rights directs attention to the fact that either self-interests and/or institutional inertia may mean that providers of services will resist changes to their provision. The argument for rights forces attention on the way providers, in resisting new changes of direction, may interpret what signals of 'dissatisfaction' they do receive from the market in ways that are conducive to their already existing production paradigm. And as a warning against elite driven reform, at least part of Jamie Oliver's problem in persuading the children to eat healthier food derived from the initial approach he took of simply deciding to offer them what *he* thought they should eat as alternatives. He had far greater success in providing alternatives once he had consulted with the children and discovered something about their already formed tastes that he could work from. Similarly, an argument based on communication rights seeks to empower subjects by making them central to the delivery of a diverse range of goods from which to develop their cultural palette. The argument for rights also poses the possibility of a conflict of interests between providers as they are presently constituted and the information needs and desires of young people.

Conclusion

This chapter has sought to sketch out the broader historical and political context within which television news about politics and young people's engagement with news and politics ought to be situated. Central to this narrative has been the mutation of capitalism under a globalising neoliberal model and its impact on young people's material and social conditions. In this context the potential of television in stimulating the cognitive awareness of young people about their society and fulfilling the communication rights of young people becomes all the more pressing. But what is depressing is the institutional inertia that must be overcome to even begin to make television more responsive to the needs of its audience. Politics has been profoundly defined by television and the media generally as forged in the triangle between Downing Street, Whitehall and the Houses of Parliament, although there is some angst that coverage of the Houses appears to be slipping down the priorities of the media as senior politicians increasingly bypass the House of Commons, for example by making important stage managed announcements elsewhere. Nevertheless, the decline of the House of Parliament in news coverage has not broken the institutionalised symbiotic relationship between the political elites (increasingly 'personalised' and 'presidentialised' by the media) and the media elites (e.g. the Westminster Lobby system) which is evident on our screens everyday, in who gets to speak, what the topics are, who the movers and shakers are, the very *mise-en-scene* of the action, the iconic settings, and so on. Most of the energy and resources of the media are focused on the elites within the political elites as the makers and primary definers of news. Yet this monopolisation of politics by the politicians (akin to the monopolisation of history by Great Individuals) is increasingly dislocated from reality. Riddell argues that the media must explore the role of the 'new centres of power... what the judges, the utility regulators and the Monetary Policy Committee of the Bank of England do' (1998: 17). Now, while this may not constitute a thrilling agenda for young people (although *how* such topics are covered is all important), the sentiment to find out the political implications and effects of what other people do, other than politicians, is important.

The highly centralised conception of politics common in the media and narrowness of conception in what constitutes political activity, produces politics as something which 'others' do, others that are increasingly remote. As Graber notes in her study of American television news provision:

> Part of the problem of attracting audiences to political news lies in the difficulty of making such news relevant to audience members. While most Americans feel a sense of obligation to keep politically informed, and a sense of guilt when reminded about their sparse knowledge of facts and figures, most do not consider such information genuinely vital to them in their personal lives. The political world seems remote, confusing and mostly dull. (Graber 1993: 84–5)

This remoteness and dullness derives in large part from the obsession which news coverage has with the factual dimensions of the story, the who, what, where and when, to the detriment of *explanatory* and *contextual* frameworks, such as *why* and *how* which Graber found far less prevalent in her content analysis (ibid.: 80). This chimes in with Philo's (2002) study that also shows that audience interest increases significantly when such contextual and explanatory frameworks are introduced.

When it comes to more participatory news programmes, outside the news bulletins, McNair, Hibberd and Schlesinger (2003) show that it has been a hard-fought struggle to get the politicians into the television studios and have them confront and engage with ordinary members of the public, while the media themselves had to overcome their own deference towards the political class in order to facilitate access to the political class for ordinary members of the public. Clearly, such encounters as constructed by *Question Time* and *Jonathan Dimbleby* are valuable and necessary, but as we show in Chapters 6 and 7, they have severe limits.

While the crisis of disengagement from politics and news about politics is widely recognised, it is less clear that established institutions and individuals who have heavily invested in the conflation between politics and 'Westminster' are in a position to really address root causes and offer attractive solutions. Following the Kevill report (2002), the

BBC website announced that an extra £5 million was being pumped into political programming.

> In particular the aim is to increase the number of people watching Parliamentary programmes on BBC television by 20%. It is hoped that improved access to the Palace of Westminster will help revitalise and modernise coverage across all bulletins and programmes. (www.bbc.co.uk/press office/press releases/stories/2002/09_september/19/politics_initiative.shtml)

The assumption seems to be that the goal is to *mend* the broken relationship between electorate and Parliament, rather than question the idea that Parliament is the almost *exclusive* place where real politics actually happens. This is the predominant tendency of mainstream studies into the 'remoteness' of Parliament. For example, Lord Puttnam's commission for the Hansard Society, *Members Only? Parliament in the Public Eye* concluded that Parliament could reconnect with younger voters if it had a better communications strategy (e.g. a better website) and if it offered greater access to Parliament. Again, there is a dispiriting failure to really grasp the extent of the problem and the solution appears to assume, as with the BBC strategy, that the problem is that people don't really *know* all the *good work* that Parliament and Parliamentarians do. In the next chapter, we turn to the broadcasters themselves, and find out what their perception of the problem is and how they are addressing it.

3
Broadcasters' Perspectives

This chapter examines how those working in various areas of television news perceive the attitudes of young people towards politics and current affairs in general, as well as towards how these are represented on television news programmes. It also analyses the measures suggested by such journalists for encouraging young people to take a greater interest in this genre of programming. The chapter is based on a series of interviews carried out in 2007 by Peter Keighron, and the full text of these is available on the project website. Those interviewed were as follows:

Steve Anderson – BBC, executive producer *Question Time.*
Kevin Bakhurst – BBC, controller *News 24.*
Matt Barbet – presenter, *Five News,* previously *BBC London News.*
Peter Barron – BBC, editor *Newsnight.*
Mark Calvert – BSkyB, editor *Five News.*
James Dagwell – BBC presenter, *News 24, 60 Seconds, StoryFix.*
James DuBern – Current TV, director of programming.
Melanie Essex – BBC, executive editor, *News Action* and *School News Day.*
Martin Fewell – deputy editor, *Channel 4 News.*
Geoffrey Goodwin – head of BBC Teens project.
Colin Hancock – BBC, editor *World At One,* previously worked on *Newsbeat.*
David Kermode – editor *Five News.*
Rob Kirk – editorial development manager, *Sky News.*
Tim Levell – BBC, editor *Newsround,* previously worked on Action Network.
Emily Maitlis – BBC, presenter *Newsnight, StoryFix.*
Rod McKenzie – BBC, editor *Newsbeat.*
Andy Parfitt, BBC – controller Radio 1, 1Xtra, and Switch.

Stewart Purvis – former ITN chief executive, currently Professor of Television Journalism, City University.
Nick Robinson – BBC, political editor.
Sinead Rocks – BBC, editor *Newsround.*
John Ryley – head of Sky News.
Nicky Schiller – BBC, senior broadcast journalist, responsibilities include news on BBC3 and the one-minute news bulletin on BBC1 at 8pm.
Chris Shaw – Five, senior programme controller (news, current affairs and documentaries).
Jon Snow – presenter *Channel 4 News.*
Sam Taylor – BBC, editor *News 24.*
Deborah Turness – editor *ITV News.*

All of those interviewed recognised that many young people do indeed exhibit a distinct lack of interest in television news programmes, although several pointed out that this is far from being a purely recent phenomenon. Whether working for the BBC, ITV, Channel 4 or Sky, all took the view that it would be intrinsically desirable if television news could broaden its appeal to attract not only more young people but also more members of other 'disconnected' sections of the audience. Several interviewees were sceptical of the mantra that 'young people just aren't interested in politics', and argued that their alienation from Westminster politics should not be taken as a symptom of a lack of interest in politics in a broader sense. This is an interesting finding, given the institutional obsession with Westminster politics that dominates television news. On the question of whether young people should be provided with their own news programmes, or, alternatively, whether the mainstream bulletins should be made more attractive to young viewers, opinion was divided, with the latter course of action being seen by some as running the risk of alienating older viewers. On the other hand, several of those interviewed pointed out that it is not simply younger viewers who watch comparatively little television news, and, in their view, making the news more attractive to young people might also bring in other hard-to-reach sections of the population.

Politics and the news

According to Melanie Essex:

> Young people say they're not interested in politics but actually if you look at the subjects they're covering it's politics, politics, politics.

> They don't like politicians, but for at least fifteen years now there have been studies saying that. But what was interesting was that when we had our party leaders on Schools News Day [see below], everybody jumped at the opportunity of interviewing them. We also wanted them to come up with their own stories and their own agendas, and they did. And their agendas were much more serious and less celebrity-driven than we thought they would be. There was a great story in a school in Glasgow about asylum seekers. There was an asylum seeker in the school who didn't want to be identified, so it wasn't a publicity stunt, but there was a petition that had been started in the school and they did a story about the petition going around the school to keep the asylum seeker in the country; in the end the story was picked up by a reporter in Scotland who interviewed an MSP about it. There was also a really good story in Kent about a campaign for half-price bus fares for school kids, and that got picked up by local radio. There was a great piece in Stoke about city centres and how they were becoming empty shells because of the developers, and they went and interviewed a councillor. It was a really serious minded, nicely made piece. So in that sense we did get their agendas because they were asked: what do you care about, what do you think is important?

A similar point was made by Mark Calvert:

> I think increasingly, particularly amongst younger people, in a non-party political, non-Westminster way, there's an increasing awareness in issues which are political. Climate change and green issues are really big amongst our younger viewers. In the user generated content part of *Five News*, what we call *Your News* [see below], given the number of submissions and videos we get from kids going on about cycling or the environment or wasting water or whatever, the environment is the biggest single thing that they're passionate about.

Meanwhile David Kermode emphasised that engaging young people, and other hard-to-reach sections of the audience, is a matter both of choice of subject matter and of finding the appropriate presentational style:

> If you do politics the right way and you ram home its relevance to people's lives then it's about story telling. We don't do a huge

> amount of politics on *Five News*, but we have a political editor in Andy Bell who's a great storyteller. I think it's fair to say that he assumes nothing in terms of the political engagement of the viewers so he always tells a story from afresh, he never patronises but on the other hand he never assumes knowledge. I think that's absolutely the way politics, to a mainstream audience, should be told. But the idea that young people find politics boring is a bit overblown because if you look at the figures for *Question Time* on the BBC they get quite a young audience, so it's about formats and the way you do politics. It's far, far too simplistic to say young people find politics a turn off.

This brings us on to the question of how best to appeal to young people in news programmes, and the topic of whether they need their own news programmes.

Tailor-made news

Sinead Rocks clearly felt that young people did need their own programmes:

> Tailoring news for different age groups is really the way to go. Young people don't feel particularly well represented, they don't see people like themselves on the news necessarily, they don't have these issues that dominate the news as big parts of their lives. So I think if there was news that was more tailor-made for that particular age group the interest would be there.

One of the ways in which this might be done was suggested by Nicky Schiller, who was interested in how news programmes might best appeal to those more used to getting their news online:

> I'm there to give them some information but my audience is never going to tune in [to BBC3] for the news, my audience is going to find the news when it's there; I hope to give them some information but it's not an appointment to view, unlike, say, the *Six O'Clock News*, for which people tune in specifically. Younger audiences won't do that in the same way as older audiences do. Our bulletins are not even on at the same time every day, they come in at the end of a film, which can be at 8.20 or it can be on the hour, it's very variable. You know the audience is there but you also know that they're not necessarily tuning in for the news, so you have to write and target the news for

them in a different way than you would if you knew they were there just to get news.

A lot of the people we would want to target are now actually on-demand type people and they're getting their news in a different way. So we should be looking at how we provide news on-demand, whether that be on MySpace, whether it be on YouTube, whether it be what we provide on the news website. You've just got to give it to them in different ways rather than the traditional 'let's sit down for half an hour and watch the news'. Some people will make an appointment to view news but I think perhaps effort is better spent on different areas to which young people have already progressed. There's a difference between not watching news in a traditional way and not being interested in the news.

On the other hand, there are those such as Chris Shaw at Five who do not believe that news programmes should be designed specifically for young people. Shaw's views are particularly interesting in that once he believed precisely the opposite. As he explains:

> We had a programme called *The Edit*. It was very much targeted at younger viewers, it was presented by teenagers and produced by teenagers, but it had a very, very small audience of less than 200,000, which for us is not really sustainable in the long term. So it was a brave experiment but to a certain extent I felt, my God, it's such hard work trying to find the younger audience and trying to hold on to them, they're so promiscuous in their television choice and they're very quick to sniff out any condescension. So I rather veered in recent years to the view that if there are young people who want to find out what's going on in the world they basically should be treated as any other news viewer and they will find it and there's very little point in tailoring your approach to what you think a younger person wants because I think they sense it and resent it quite quickly.

Of course, in a commercially driven operation such as Five, audience figures are particularly important, and Shaw was concerned not only about being unable to attract large numbers of younger viewers but also about the danger of possibly losing older viewers by becoming too youth-oriented. As he explained:

> I tried very hard in the early days of *Five News* to do things that I thought would appeal to younger viewers and so on. We had younger

people working on and presenting the news (most famously with Kirsty Young), we deconstructed it, we took it out from behind the desk, we presented from the newsroom. We were one of the earliest examples of trying to bring down the barriers, which is the clarion call of modern digital news: remove the barriers, let the people in and so on. So we did all that and our agenda focused a lot more on protest movements and music and showbiz issues and student issues; it did have an effect but it also restricted our overall audience growth as well. It made us feel a bit too niche for a mainstream terrestrial entertainment channel.

In contrast, the publicly funded BBC, with its public service remit, can afford, and indeed is required, to take less of an either/or approach. As Nicky Schiller puts it:

On the one hand some young people say they just want short stories, headline updates, *Metro*-style, editorially free, bite-sized news. On the other hand some young people say they want more background, more in-depth stories, they want things explained more. I think they want both. Again, they are intelligent people. So sometimes all they want to know is 'has anything happened in the last hour, do I need to know anything?' That's what those 60-second bulletins are designed for. But if you want in-depth, then the BBC does a lot of in-depth stuff elsewhere. If you want your ten-minute pieces then *Newsnight* will always carry on doing those pieces and you will always be able to find that in-depth side of things. There may be an argument that we need to do more in-depth things for a younger audience. News 24 has been experimenting with Teen 24 and there are different projects in the BBC to try to do more things of depth and in a long form for a younger audience, but you only have to look at the viewing figures and the demographic of *Panorama* now that it's moved into primetime and is on for half an hour and in a slightly different form, to see that it is getting a younger audience than it has done in the past. There is an audience for both, I don't think it's one or the other.

Or as Emily Maitlis argued:

I think it's fair enough on *Newsnight* to be interested in questions like House of Lords reform, and I think it's absolutely acceptable for *Newsbeat* to leave that permanently off the menu. It all comes down

> to variety of outlets and variety of programmes and I think that's going to intensify as opposed to diminish.

Maitlis' point shifts the discussion from the formal qualities of news programmes to consideration of their actual contents. On this matter, Nick Robinson was refreshingly forthright, stating that

> We should try obsessively to engage people in news but the obvious point to make is that young people (but this is true of large parts of the audience as well) are not interested in political process for its own sake. They're not interested, per se, in the internal workings of parties, in who's up and who's down, the Westminster village, but they *are* interested in issues. So the obsession, since I've worked in news, has been to try to make things about issues, about people's lives. So at a crude level you would not now get what you would have got fifteen years ago when the news would have said: 'The Conservatives attacked the Labour Party today over a Select Committee Report about whether schools should select'. Now it would say: 'There's an argument going on about whether *your* child in *your* school should be selected in a new way'. And then the Westminster politics would come at the end. So there's already been a huge sea change in trying to see stories in their top line and their treatment through the eyes of the audience.

Robinson is dismissive both of purely cosmetic changes to the news and of the notion that young people are not interested in political issues, and although he is sceptical that there is any single answer to the problem of young people's (and others') disconnection from the news, he is adamant that changes must indeed be made both to the content of news programmes and to their presentational style if they are to attract wider audiences:

> Clearly, you have to ask yourselves each of these questions: Is it relevant to people's lives? Are you putting the issue before you put the process? Are the people presenting it 'them' rather than 'us'; are they white, middle-class, middle-aged blokes in suits? Are they using too much jargon? All these things are obviously potential barriers to people's engagement with the news and arguably one could do more at every stage about breaking down some of these things.

Another barrier to engagement with news programmes is identified by Robinson as the all too frequent assumption of knowledge, an area in which he clearly believes that programmes aimed at younger viewers and listeners have much to teach more mainstream news programmes:

> One of the things that I think is common to *Newsbeat, Newsround* and actually Radio Five Live, which I've worked for as well, is no assumption of knowledge. They constantly seek ways to simplify without being patronising, to go back to basics and say we know you're starting from scratch here, to tell the story from the most basic, to tell it from the perspective of the people who are new to it, not to see it through the prism of people in power. Those things seem to me palpably popular and they work. Quite a lot of that has been built into the BBC and ITV early evening news programmes, but you can do more of it.
>
> If it were my job to make the news more attractive to younger viewers I'd go to things that already work, things like *Newsbeat* and *Newsround* and Five Live, and really micro-analyse what do they do with that story and not that one and show it to groups of young people and say 'what do you prefer'? The other thing, which everyone always says but finds very difficult to do, is that you've got to change the intake of people making news, we're all very similar. We're all horribly similar.

'Yoof' TV

Although those interviewed agreed that television news aimed specifically at young people needed to adopt a different format and style from mainstream news programmes, they were adamant that the young audience should not be talked down to or patronised. As Sam Taylor put it: 'the project with the big "It's Youth News" label on it has got the kiss of death from the start', whilst Sinead Rocks warned that 'you don't want to turn into some hideous "yoof" TV which is overly aware of itself in a way that would be a switch off for younger viewers', and Emily Maitlis felt that

> A programme that started, 'hey kids, you're really going to love this one' would be more of a turnoff than a fairly grown up delivery of what the newsroom considers to be the top stories of the day. I think that talking down or 'trendification' could backfire quite badly... If somebody said to me 'appeal to the youth market', I would throw

> something at them. That for me is 'hey kids', and the moment I say 'hey kids' fire me or shoot me, or both.

In this respect, Tim Levell told a revealing story about the former BBC3 programme *Liquid News*:

> It was quite sound bitey and had lots of showbiz news, and a lot of stuff with a 'we don't really care' attitude. Quite famously, on the very first day they launched they had on their website this talking point saying 'give us your feedback about our show' – and they called it 'Shite or Alright?' That was meant to be their kind of edgy thing. And it bombed, and one of the reasons was that when young people decided they wanted news they wanted straight news, they wanted proper news done intelligently and done well when they were in the mood for it. What they didn't want was someone to try and catch them when they weren't in the mood and try to give it to them in a different way.

Interviewees also agreed that there was no one single, successful recipe for making young people's news programmes. Sinead Rocks argued that 'to assume that young people can only take short, sharp bursts of information is slightly patronising. I think it's about offering choice and diversity and not alighting on one formula and thinking that the one size fits all approach is going to work', and Rod McKenzie added that

> It's a real mistake if we think that if we hire this presenter or if we get this writer to write the news in a certain way, or if we shoot the news with all our camera crews lying flat on the floor and shooting upwards, that'll be the magic bullet that means millions of young people will suddenly think the BBC is super-cool and worth watching. Because actually there are dangers about being super-cool, because a lot of young people don't want super-cool, they want familiarity and comfort.

A not dissimilar point emerged from the interview with Mark Calvert, who reported that

> From all the focus groups I've been involved with and research I've seen about this I think that even amongst young people there's a genuine hunger for information – let's call it broadly information

> as opposed to what we as editors call news – and young people are probably consuming more information than they ever have. And so we need to open ourselves up first of all to a wider definition of what constitutes news to a younger audience. I think that's quite key. Secondly, we have to recognise that actually, even if they consume news in different ways and have a broader view of what a news story is, there's still amongst younger people something about news that they don't want mucking around with too much. There's something about news – and perversely it goes against the grain of appealing to youth – that's got to be proper, that's got to be right; they actually want to have a bit of authority and they don't want it messing around with too much... I've spoken to some of the focus groups that have looked at things like *60 Seconds* on BBC3 and even though they like brevity and they often refer to the *Five News* updates more than they do to our full programmes, they don't like it if there's a box with one lot of pictures and a presenter over here and all this graphics stuff. What they tell me is that it's like their dad trying to be funky. They just want to be told what the three or four top stories are very briefly and then if they want more they can go and search it out. That to me underlines this idea that for them there's something about news that has got to be proper.

Similarly, Kevin Bakhurst stated that, on the basis of his discussions with younger audiences, he had come to the conclusion that 'they don't necessarily want to turn on BBC news to get the latest celeb news. They've got lots of other sources for that and they still appreciate the fact that we do serious news'. Specifically on the question of presenters, Melanie Essex noted young people's marked preference for the not-particularly-young George Alagiah and concluded that 'it's just that some have got the chemistry and some haven't', whilst Emily Maitlis argued that

> I don't think young people have a problem with, say, Larry King, who is a middle-aged, white, grey haired man in a suit, or Jay Leno or Jerry Springer. Nobody sits there going 'what a boring old man in a suit'. The right man in the right frame of mind gets a huge teen audience. There is a buttoned up approach that sometimes accompanies that sense of a man in a suit but I don't think young people are that superficial. She also added that 'it's a lot more than just that; I think it's about the way you choose your stories, I think it's about the terms you use'.

We shall return to the question of presenters in the section of this chapter devoted to Channel 4.

Involving young people in production

Discussing the question of whether news programmes for younger audiences should have a younger feel to them, Sinead Rocks argued that any decisions on this matter 'should come out of a lot of concentrated audience research and focus rather than having some people in W12 deciding what young people must be into'. Her experience of discussing these issues with young people had led her to conclude that what is important is to

> Provide a forum for young people to make the news themselves, dictate the agenda and then influence how the news is treated and presented. I think we're dealing with a very media savvy and media literate age group and frankly some of the way news is presented bores them. They can go out and make their own films and tell their own stories and I think news needs to move to a place where it becomes much more of a process rather than us swooping on someone's doorstep and doing it all ourselves.

Here the BBC News School Report project has played an important role, helping 11- to 14-year-olds to make their own news programmes which are broadcast on their school's website on an annual News Day in March. On the day itself, a special School Report channel is available via the red button, and extracts from some of the programmes are shown on mainstream channels, particularly *BBC News 24*.

Such considerations clearly bring us into the somewhat fraught realm of user-generated content (UGC), and it is interesting that two of those interviewed expressed reservations, albeit very different ones, about this kind of material. Sam Taylor pointed out that

> It's always been the case with programming aimed at young people and children that the danger is that you end up with a bunch of well behaved, high achieving, white kids from the Home Counties. If UGC is a way to get young people more involved in producing and editing their own material, how do you do that in a way that is not the equivalent of interviewing an expert from an esteemed think tank? Because you're cherry picking people with the access to the kit and the information. This is another justification of the continued

role of the producer which, as a career producer, I must always try to avoid falling into, but there has got to be a level of assistance to make sure that you help some people and move their agenda on.

However, Melanie Essex made a quite different point about UGC:

I've got a colleague whose fourteen-year-old son says he and his colleagues spend the whole time making films with their mobile phones. They interview each other and do other stuff. Now I've got a twelve-year-old daughter and she would no more do that than fly to the moon. She and her friends just don't do that. So we're making these huge assumptions that there are lots of people out there who are going to be making user generated content, but I think we're actually a little way off it happening. I think it's really important that we can say to people, though, and that's what School Report is all about, that there are opportunities and we're interested in what you've got to say and we have a showcase for it. But I think you've got to signpost it clearly and set the boundaries clearly.

Five News now has a UGC slot entitled *Your News*, most of whose contributors are under 25. And in spite of his earlier reservations about news specifically tailored to younger people's tastes, Chris Shaw is positive about this development, noting that

They [young people] are just much better at it than older people, technically, but also they're just more imaginative about what they do. There's some evidence that that kind of YouTube/MySpace phenomenon might in itself generate a kind of newfound interest in information and campaigning...I do feel that there [in new media] lies the answer. The people who send in stuff and want to go out there and become citizen journalists tend to be in their teens or early 20s. They may be a very small sub-group of young people but they are young people nonetheless and the evidence is that it's quite compelling for younger people to feel that they can do it themselves without someone over there telling them what's going on; they like to feel that they are in some sense participants, and we aim to build on that. What we're trying to do at the moment is to try and treat news as a kind of club that you belong to. I know it sounds ridiculous but we're trying to engage with our audience much more intimately in what we do and to make them feel participants. It's true that younger people are more inclined to follow show business

> or sport stories or music stories or eco stories but I think that they're much more interested if they feel they can be part of something or if they're part of a community of interest or if there's an issue they can pin their colours to.

Martin Fewell, deputy editor of *Channel 4 News*, also drew attention to an interesting project aimed at engaging a younger audience in the news by involving them in actual news production. This was Breaking the News, a media literacy project which was the result of a collaboration between Channel 4, the British Film Institute and the BBC aimed at raising awareness in 16- to 19-year-olds of how news is made. As Fewell explained:

> We brought everybody in to the ITN studios over several months in a process which involved us tutoring and mentoring people from ten different schools, then getting them in here to make news for a day, and then putting the product on our web site and also reporting it on the programme as well. And that was part of an understanding that government and public service broadcasters share, a generalised concern about people's understanding, particularly young people's understanding of the process, and trying to engage in it.

Channel 4 News

As we will note in Chapter 8, *Channel 4 News* is particularly liked by a significant number of younger people. It is therefore instructive to examine what broadcast journalists believe its appeal to the young to be.

Sinead Rocks refers to *Channel 4 News* as

> Like *Newsround* grown up because they really think about context and levels of assumed knowledge and they do explanations incredibly well. They also have a smaller pool of on-screen faces so you actually feel you can build up some level of knowledge of them, more familiarity with them, whereas the BBC is such a huge machine that you don't necessarily get to build that kind of ... bond is too strong a word, but that kind of relationship. It looks very different, it gives more understanding and it seems to go below, do more than scratch the surface, getting the background and debating the various points. And it's also got a very charismatic front man in Jon Snow, and his appeal probably has a lot to do with it as well.

Martin Fewell, deputy editor of *Channel 4 News*, usefully explains the programme's distinctive features by reference to Channel 4's overall remit:

> We've always been very conscious of the fact that we are the news service for Channel 4 and that means, on one level, making sure that what we do corresponds to the values that Channel 4 espouses. We take Channel 4's values as the foundation of what we stand for but then we try and translate those into slightly more specific *Channel 4 News* values and objectives. Those are in many ways similar, about challenging consensus, representing minority, diverse communities that might not otherwise appear in mainstream news, picking up on popular points of view and challenging those in powers and authority, whether that's political power or economic power, and asking difficult and awkward questions, taking particular interest in international development stories (foreign news is a high proportion of what we cover) and not being worried about doing stories that viewers might find difficult or that not many viewers might want to watch. Ratings are not our main concern in life; if they were, Channel 4 would not be running the news in the slot it does.

This rubric, in Fewell's view, is one that clearly touches a chord with the younger audience:

> I'd say the attitude of the show is important for that kind of audience – that it is challenging, it's not seen as being a mouthpiece for established consensual thinking, it's prepared to tackle difficult subjects and subjects that younger viewers are interested in. It's long had a particular interest in the environment and climate change and in the last two or three years it's had an obsessive interest in that area. It also tackles subjects like trade and the developing world, which some young people are interested in. I think the overall look and feel of the show is designed to be contemporary and modern and although it's obviously a fact that Jon Snow is one of the older television broadcasters there is it's also a fact that he clearly has an appeal to younger viewers. That's not something that he or we deliberately cultivate but it just grows out of his personality and his character and his approach to his subject. He has a very youthful energy in the way that he addresses a subject. So I think there's a whole range of things in there but what I'd also say is that we're not trying to deliberately tick the subjects that we think the younger

> age group is interested in. Our fundamental brief is about tackling serious news in interesting and provocative ways, challenging consensus but doing so in a way that's creative and imaginative and with high production values. We don't specifically go out of our way to cultivate a particular age group within the audience, our duty is to give people the kind of news that we think they don't always get elsewhere, and the proportion of our audience who are younger proves that, for at least a certain group of people, there is an appetite for that kind of news.

Unsurprisingly, engaging young people in television news is something about which Jon Snow cares passionately, and he is characteristically forthright both in his advocacy of the programme's news values and his critique of those of mainstream television news programmes:

> The programme is not aiming specifically at youth, and anyway I think aiming at youth can be patronising – youth don't want to be aimed at. Traditionally stuff that is aimed at youth has failed. We aim at what is going on but if people are talking crap we'll try and expose them as talking crap, and that's what some young people like, seeing somebody mauled for being an idiot. It means never taking anything for granted, it means never accepting an official view of things.
>
> I think we're much more anxious than other broadcasters to get young people talking with their own voices rather than going in to shape a piece to fit a perceived development. I think one would want to test a development before one preceded to build a whole report around it. We'd want to ask: is it true what they're saying or is it just a bit of sensationalism? I think we try to empower people when we meet the community; we ask them: what do you want to say about this? You live here, what do you make of it? Rather than doing what they do at ITV which is to say: go down to Peckham, do a piece to camera at the bottom of the stairs, talk to the copper who'll be there, then talk to someone from the community and then do a little sign-off at the end standing on Waterloo Bridge. Well, forget it. What do they know? Maybe the cop's an arsehole. What's the point of predetermining what you're going to do? It's the old John Birt phenomenon, where you predetermine what the documentary or the news report is going to say and then you make the facts fit what you're going to shoot. No, I say go down there, find out what these people want, empower them, let them talk, let's hear them. And the beauty

of *Channel 4 News* is we've got the time. Time is the essence. If you're going to do it in one-and-a-half minutes it's liable to be the John Birt formula. If you're going to do it as we would, we might devote fifteen minutes to it, and fifteen minutes is a lifetime in factual television.

BBC news provision

As the largest public service broadcaster within the UK, the BBC has a special obligation to address the disconnection between its services and young people. It has conducted numerous reviews and initiatives in recent years that have identified the need to connect with young people more successfully across a whole range of programme provision, to make the BBC relevant to the lives of young people. News and current affairs programming is recognised as a key area that needs to do better with young audiences. As James Dagwell (*BBC News 24*) puts it, rather brutally:

> ...the people who watch the 1 o'clock news are going to be dead soon and where's the audience going to come from? Ultimately you could argue that that's going to threaten the funding we get and of course what goes on in the back of management minds is that why should they plough money into television news when the audience is shrinking and getting older?

The proliferation of choices in terms of television viewing means that the era when young people may have watched television news (and begun to acquire what broadcasters call the television 'news habit') while their parents were watching it is probably over. As Rod McKenzie, editor of *Newsbeat* on Radio 1 puts it

> I think what's changed with young people is the easy availability of alternatives. My Space, BeBo, Face Party, all that stuff that's going on out there means there's a lot you can do rather than listen to the radio or watch telly. There's also itunes and ipods, YouTube, where do you end? There's a lot of stuff that you can watch and listen to. Now there's also a lot of evidence that for young people news is either ultra local, underground or conspiracy theoryish, so it's a lot more entertaining, if you like, to look at a film on YouTube or follow some links on the web to something which says that the government's doing something dodgy or whatever it is rather than get into what your parents would watch, which is probably the BBC.

The problem for the BBC is that the main news bulletins on BBC1, the 1.00pm news, the 6.00pm news and the 10.00 pm news, all have an ageing audience profile of people in their fifties and sixties as their core audience in terms of both size and loyalty. While this audience sustains the news, it also threatens its long-term viability. The dilemma though is that there is a danger, as the broadcasters see it, of alienating that core audience if they seek to appeal to young(er) audiences (here younger, may mean people in their thirties and even forties, rather than really young audiences, in their teens and early twenties).

Because of the difficulties of experimenting and innovating within the main BBC1 bulletins, new ideas and trends, such as they are, have tended to be explored in other parts of the BBC system, with the idea of creating synergies amongst the different programmes. *BBC News 24,* for example, has been one important area where journalistic cultures within the organisation have been modified. This is because *News 24* appears to fit in rather better with younger audiences' television consumption habits, being in effect a service they get something out of 'on demand' rather than having to make an 'appointment to view' at a set time, as with the main BBC1 news bulletins. According to Sam Taylor, 25–34-year-olds were, by a small margin, the biggest audience, as defined by age, for *BBC News 24* in 2006. There is, however, a strong temptation to interpret a younger news agenda in terms of market led stories that do little for addressing young people's needs as citizens. Taylor calls 'resonant entertainment stories' a key plank of *News 24*'s attractiveness to younger audiences. Thus James Dagwell from *BBC News 24* tells us of an entertainment strand he has started, twice nightly on the channel, admitting that it is 'quite celebrity-based news', with the aim that it helps 'the image of the channel'.

> Two or three years ago they would never have put a ten-minute entertainment slot on News 24 at 6.30 in the evening because most people in the building would say we don't do that on News 24. There are still some people who say that but my argument is we absolutely should be doing it, we do sport, we do business, be brave enough to devote ten-minutes of an hour to stuff that other people are interested in. The whole explosion in Heat magazine style stories, whether they're gossipy or not as long as they're founded in some sort of truth, why not? Why shouldn't we be doing them? I think it's really important that we're brave enough to do that as well as the harder stories that the BBC has always done and is well known for. We've got to start

> branching out and touching all bases and covering all genres of story.

As a radio broadcaster as well as a television broadcaster, the BBC is in a position to look at other branches of its provision and see what works that might be transferred to television. One possible model for a news programme that could speak to young people is Radio 1's *Newsbeat*. This 15-minute news service sits within a radio channel that has not only a young audience, but also attracts working-class audiences. *Newsbeat* reflects this audience composition, just as Radio 4's *World At One* news programme reflects its older and more middle-class audience profile. For Rod McKenzie, editor of *Newsbeat*, the audience composition for his programme, quite unusual in news terms, means 'doing politics' in a different way. The established institutions of political business are seen as problematic from the point of view of engaging the audience:

> Parliament has a bad image, it's lots of mainly men in suits, certainly all middle class, elderly people, or middle aged people from a younger audience's view point talking about quite boring stuff in an inaccessible way, and that's a switch off.

Avoiding the default locations and personalities of political coverage (MPs, Houses of Commons, 10 Downing Street, etc.) *Newsbeat* often does 'politics without the politicians'.

> So what we try and do is do that story but told through case histories or real people's experiences. So, for example, if it's a hospital closure story we would try and find somebody whose treatment has been effected by a hospital closure or would be affected by a closure, real people, real stories. Then people say, hang on a sec, you don't give the politicians a chance to have their say.
>
> Rod McKenzie: The politicians.

Along with this different approach, *Newsbeat* also has younger reporters, while radio stations such as 1Xtra (Radio 1's black sister station) and Asian Network have demonstrated the importance of having reporters from the same communities as those in the news, in order to access hard-to-reach groups and get them to articulate their perspectives on a story. Sam Taylor from *BBC News 24* has been working to try and build up the channel's network of contacts with young people. This requires getting journalists to rethink established practices, where they reach first for well-established organisations and lobby groups whose job it is to put people up for interviews.

> ... it's another one of these things where our overall approach can shift, and that often benefits the younger viewers, if we put more focus on case studies and user generated content and what people themselves are saying about stories. We should make our content more accessible and more interesting. So that's one of the things we do, we'll follow up people who get in contact with us through text or email to use them as interviewees or case studies and things like that.

And Taylor also sees such approaches cross-fertilising over to BBC1:

> The 6 o'clock news has done pieces recently, for instance the one about young black men needing better role models. 6 did a piece totally centred in the community talking to young black people about who their role models were, whether they needed them at all, which ones wouldn't they prefer and actually trying to unpick that. I think that in the end is a much better piece of coverage than where that story usually starts off, which is where a kind of august body makes some recommendations that this should happen.

News values and negativity

As we explore in more detail in Chapter 5, the image of young people communicated by broadcast news is all too often a negative one, and the evidence suggests that this is a significant cause of many young people's antipathy towards news programmes – and indeed towards the news in general.

Of course, as all studies of news values make clear, stories with negative overtones traditionally rank high on news agendas. From this perspective, stories about young people and knife crime, for example, could be seen not as an example of young people being specifically singled out for negative treatment by the media but simply as a consequence of journalism's habitual gravitation towards 'bad news' stories. As Mark Calvert puts it:

> It's not just confined to younger people. Most news by its nature is bad news, we're largely reflecting the grim stuff that happens in life

> and the serious stuff that happens in life, and so we will see middle-aged people or old people who aren't reflected in a particularly great light because they're involved in grim, bad stuff. So I don't think that's specific to young people at all.

Similarly, Deborah Turness argues that 'most stories in the news are stories about bad things that have happened, so most stories about kids are going to be about bad things that have happened to kids' whilst Kevin Bakhurst states: 'a balanced portrayal of young people is important. But news generally tends to be more negative than positive and probably kids get the bums rush on that as well. But that criticism is aimed at us in almost every area, that we only cover things that go wrong and it's negative'.

At the same time, however, most of those interviewed were well aware of the pitfalls of such an approach to news (whether about young people or other sections of the population) and sympathetic to attempts to change it. For example, Sam Taylor was critical of the relentless information flows which can all too easily push certain kinds of stories, along with certain kinds of routine, taken-for-granted ways of dealing with them, onto the news agenda:

> Young people in this country are not exactly marginalised, but a storyline of youth in Britain is usually not enormously positive. A lot of our thinking is trying to change that, but you don't always see it on the surface because the stories keep coming at you all the time. We're trying quite hard to open up space but on a bad day, the publicity machine, if you don't watch out, will deliver for you a story like A-level results or changes to university admissions procedures. You could quite easily sit there all day interviewing people over the age of 40 about what they thought about it. You could go around all the political parties, they'd all have a view on it, and you could then get someone from admissions a vice-chancellor, and so on. So a lot of what we're trying to do involves saying that there's a limit to that kind of approach and that we've got to get real case studies, we've got to get younger people who have been through these situations or are working through them now.

A similar point was made by Nicky Schiller:

> At some point you have to say, right, let's get on the people that this actually affects and get them talking about it. With school meals and

> things like that, too often it's adults talking about the subject rather than getting kids actually to do it. At BBC3 we tried really hard if there was a story about turkey twizzlers or something similar to have a ten-year-old in the studio telling us what they thought rather than having an adult telling us what they thought children thought.

This indicates an awareness that the 'good news vs bad news' dichotomy does not capture all the significant issues involved. Irrespective of an evaluation as to whether some news item is good or bad, positive or negative, there is the issue of involving young people themselves, and beyond that, giving them some power over setting the agenda. Such an approach returns us to the importance of involving younger people in the television news production process itself. As Rob Kirk put it when asked if *Sky News* concentrated too heavily on negative issues concerning young people:

> You can't turn your back on these issues, but often when we're dealing with them we try to get young people to take part in the programmes or even to get young people to do their own reports, which is quite a common technique with us. It's always good to get a young perspective on an issue, particularly if it relates to young people. We've done quite a few reports, for example, on kids and diet. We've also got young people to produce their own reports. We had a series last year called 'Little Monsters' which was about young people's behaviour and we got young people to do those reports. So without patronising young people you can bring them in and get them to engage and to do their own reporting.

However, Melanie Essex took the argument a stage further by stressing the importance of involving young people in a wide range of stories, and not simply those about their age group:

> Teenagers aren't necessarily a group to whom we would go to get their perspective on other things. However, there are people within TV news who are thinking about this and about representing them a lot more. So, for example, if we were doing a story on global warming it wouldn't necessarily occur to us to go to a bunch of sixteen-year-olds and say: what do you think? Actually it might do now because we're trying harder, but a couple of years ago we'd have gone to politicians and we'd have done it from that perspective. I think it's about mindsets and quite a bit of work's being done, certainly

in the TV newsroom and in News 24, to try to broaden the range of voices and faces on television. In fact people have been trying to do that for as long as I've been in news but there is a specific focus at the moment in getting on younger voices and faces in more neutral stories rather than just the hoodie/knife crime/truanting/teenage pregnancy/anorexia stories.

Conclusion

What these interviews clearly demonstrate is that within the main institutions of broadcasting, many journalists are giving a great deal of thought to the question of how better to connect young people (along with other alienated sections of the audience) with television news, both in its mainstream and youth-oriented forms. We can summarise some of the key themes to have emerged from this selection of the interview data:

- There is a general recognition that politicians and Westminster politics in particular is a turn-off for many viewers, including young people. And yet the implications of this insight have yet to feed into any real innovations in news provision, at least on television. Attempts to address this issue at all have usually been limited by the aim of trying to make Westminster politics more attractive to the public.
- It is widely recognised that trying to make news more attractive to young people by ramping up the amount of celebrity or entertainment led news stories is patronising and simplistic. And yet at the same time, this is often precisely the kind of strategy that broadcasters reach for in practice.
- It is widely recognised that the relationship between television news and its audience is strongly mediated by the presenters of broadcast television news. Significant factors here are the language they use, the location(s) they are situated in (especially the designed location of the news studio, but also the external locations they visit), and the representativeness of the presenters in terms of race, gender and class. While there has been some progress in terms of race and gender diversity amongst presenters, class remains a particularly intractable problem in such a 'professionalised' world as news journalism.
- There are opinions expressed on the necessity to have both a universal news and a tailor-made news provision, although not everybody agrees that both are necessary. But realising both is extremely difficult. The mainstream news provision is notionally, but not really

in any authentic way, universal in terms of who is included and who it is primarily aimed at. But there are also real problems in trying to develop a tailor-made news programme for young people, both in production terms (avoiding the pitfalls of the dreaded 'yoof TV') and in terms of sustaining anything tailor-made within the schedules that are hideously competitive. For this reason, many young people in their late teens continue to watch *Newsround*.

- Involving young people and indeed ordinary people in general in the production of television news opens up real possibilities. But it also raises a host of questions. What is the role of the professional producer when access is opened up? To what extend do they continue to set the agenda? How far does the involvement of 'laypeople' in the production process allow for an expansion of the repertoire of approaches to and understanding of the news agenda? How far is it an opportunity for the broadcasters to extent their influence and legitimacy by engaging in a bit of token inclusiveness? How far do ordinary people end up internalising the values of the 'news machine'? (We will address some of these issues in Chapter 5).
- The relative success of *Channel 4 News* in attracting young people suggests that there is some truth in the truism that a more attitudinally sceptical approach to dominant institutions and news agendas pays dividends with this audience. But how far are broadcasting institutions willing to defend such approaches in practice? How far can such approaches be taken?
- There is a widespread recognition of the need to change journalistic cultures, especially vis-à-vis their routine sources, the people they speak to, the people they gather evidence from and the people they recruit as interviewees. There is also some recognition that young people need to be included more across a wider range of news topics, and not just included in stories that involve them directly (although even that is not yet routine). However, changing ingrained cultures is particularly hard, and initiatives that are flavour of the month can quickly disappear leaving the main edifice of habitual practices intact and untouched.

Subsequent chapters in this book will make clear that there is indeed a long way to go before some of the better insights expressed here by some individual broadcasters are translated into institutional reform and new more relevant news provision.

4
Content Analysis of Television News

Thus far we have situated the disconnection between young people, television news and politics in a number of contexts. The broadest context in which the problem can be situated is that of the relationship between news organisations, the state and the socio-economic relations of capitalism. This broader context, particularly the fundamental division between political and economic power, the subordination of political power to economic power in the era of neoliberal capitalism and the particular disenfranchisement and vulnerability of young people in this context, must be borne in mind as the essential background to any understanding of what is happening in the field of television news and its relationship to young audiences. But obviously we must move in closer to the institutions of the news broadcasters themselves to get a more detailed account of how the problems generated by these large-scale structures are conceived/perceived by broadcasters themselves and how they play out in relation to the question of young people's disconnection from news and current affairs output.

At the more specific contextual level of the broadcasters themselves, we have explored some of their verbal responses and journalistic practices and strategies that have tried to address what they acknowledge as a problem of disconnection amongst young people in relation to their news output. In this chapter we begin a first sweep through the actual output of the broadcasters, which formed the main sample for our quantitative and qualitative analysis. Whatever awareness there may be that there is a problem and whatever strategies are currently being pursued with a view to modifying news and current affairs output, we need to understand the nature of that output as it presently is.

In this chapter then we sketch out *what* television news focused on during our sample period, through an overview of news topics, and *whom* it paid most attention to, through an overview of the relative prominence of different social and political 'actors'. When interpreting and discussing these results we give particular consideration to the extent to which broadcasters provide a news service that is interesting and appealing to younger viewers. We also look closely at how well broadcasters manage to *include* young persons, and 'the public' more broadly, in news discourse.

News consumption and political engagement

It is commonly argued that consumption of news is positively correlated with political participation (e.g. Hooghe 2002; Norris 1996; Newton 1999; Pasek et al. 2006; Putnam 2000). As Besley has summarised, 'despite a number of writings decrying the negativity and corrosive content present in much news content ... studies strongly support the idea that attention and exposure to news leads to higher levels of political participation and participation in the community' (2006: 43). Putnam (2000) found, for instance, that people who consume a lot of news often are involved in political and civic activities, while those who consume little news tend to participate significantly less.

Another common hypothesis is that the more knowledge and understanding one has of politics, the more likely it is for one to become interested and engaged with it. White, Bruce and Ritchie suggest that this is one of the main reasons behind young people's lack of political engagement. They characterise young people as having 'depressingly low levels of political interest and knowledge' and suggest that young people's 'lack of knowledge and understanding about politics, and the difficulties they perceive in trying to grasp such a "complex" and "dull" subject, leave them with insufficient access to political matters' (White, Bruce and Ritchie 2000: vi). Consequently many simply tune out.

For such reasons it is widely agreed that one key to increasing young people's engagement with politics is to increase their consumption of news. At the very least regular news consumption has the potential to increase awareness, knowledge and understanding of politics and public affairs (even when we acknowledge, as we must, the limited views and interpretations on offer within television news). Yet, as outlined in the introduction, young people consume significantly less news and current affairs than their elders and it is in fact declining. As part of a solution to political disconnection it is therefore important that broadcasters deliver a news service that is more appealing to young demographics.

Central to this question is, of course, establishing what young people find interesting and will want to watch.

Young people and 'soft news'

One of the most common assumptions on this matter is that young people are mostly interested in 'soft' news topics such as sport, consumerism, entertainment and lifestyle themes. As Buckingham outlines, 'young people's use of (and interest in) news media is minimal. Only 6 per cent of young people's TV viewing comes into this category, while their newspaper reading focuses largely on entertainment, features, and sports pages' (Buckingham 2000a: 251). For this reason Buckingham is sympathetic to 'popular' journalism forms, which he believes can better engage the youth audience.

> Far from being mere sensationalism, the emergence of more popular forms of news journalism – such as tabloid television and 'faction' shows – could be seen as an attempt to engage more fully with the changing orientations of the younger audience. (Buckingham, 2000a: 251)

According to this viewpoint, one way of attracting younger viewers is to devote more time and space to soft news.

Assuming increased soft news will in fact attract more young viewers, the question needs to be asked as to whether consumption of 'tabloidised' news will actually increase young people's political knowledge and engagement. Critics of 'tabloidisation' believe rather that it will lead to a 'dumbing down' of public communication and ultimately to a weaker democracy (Blumler and Gurevitch 1995; Franklin 2004; Patterson 2000). These critics also fear that focusing on 'trivial' matters may distract citizens from 'important' issues and undermine the sense that politics is something one should make an effort to learn about. From this perspective, tabloidisation and increased levels of soft news will merely contribute to a less informed, less engaged citizenry.

Not everyone shares this view, however. Even critics of tabloidisation tend to accept that evolving audience preferences are one of the driving forces behind tabloidisation. Indeed, the Pew Research Center in the USA believes the pursuit of young viewers and readers specifically has been a central influence. According to the Pew Research Center (1990: 2), young people's preference for 'lighter news' has 'buoyed the popularity of the new lighter media forms' and influenced 'even some more traditional media' to adopt 'a softer news focus in response to diminishing levels of interest in hard news'.

For this reason some scholars have welcomed the development, viewing it as a democratising trend in media provision. Baum (2003) and Graber (2003) believe moreover that it has increased the appeal of news programming to viewers who are less interested in 'hard' news, meaning that such viewers now consume more news about politics and public affairs than they did before. Proponents of this viewpoint therefore reject the 'dumbing down' thesis, arguing that the net effect of tabloidisation is, paradoxically, that more people are exposed to news about politics and public affairs, leading to greater public awareness overall.

An obvious caveat to the pro-tabloidisation perspective is that news bulletins should not be completely overrun by 'infotainment' and should carry enough serious content for viewers to learn something worthwhile by watching them (Brants 1998: 329). Replacing the BBC's *Ten O'Clock News* with *Entertainment Tonight* would, for instance, do little good for British democracy. On the back of the tabloidisation debate, a number of quantitative, longitudinal analyses have measured the levels of 'hard' versus 'tabloid' content on television news, comparing contemporaneous news bulletins to samples from decades earlier. These studies did find significant levels of soft news and confirmed that tabloid topics are more prevalent today than previously. An analysis by Barnett, Seymour and Gabor of UK TV news between 1975 and 1999 reported that 'there has undoubtedly been a shift in most bulletins towards a more tabloid domestic agenda' and 'there is indeed a degenerative process of "dumbing down" in television news' (2000: 12). Similar research by Winston compared research data from 1975 with a sample in 2001 and concluded that tabloid content had risen significantly over the years (2002: 19). However, both studies revealed that hard news topics continued to predominate overall. Ultimately, Barnett et al. concluded that British television news continues to provide 'a healthy balance of serious, light and international coverage' (Barnett, Seymour and Gabor 2000: 12). Hargreaves and Thomas reached broadly similar conclusions in an analysis of content from 2002, and found between programmes 'a considerable degree of diversity in contemporary television news content' (2002: 88).

Young people and 'hard news'

Contradicting the finding that young people prefer soft news is research showing that they are also easily engaged with 'hard' news topics. As we discussed earlier, while younger viewers may not be particularly interested in news about formal politics and politicians, they tend to be much more responsive to news about issues. Henna and Weinstein found that young people's disinterest in political news stemmed primarily from a distaste for partisan bickering and point scoring, dishonesty, and inaccessibility, at Westminster (2006: 522).

It is also commonly suggested that young people are disproportionately active within 'single issue' campaigns for causes such as the environment, human rights and animal protection (Hall and Williamson 1999; Power Report 2006). As Bentley and Oakley outline,

> younger people are more likely to be engaged in activities such as single issue campaigning and newer political causes which established political parties still struggle to adapt to. There is evidence that younger people are more likely than many assume to sign petitions, engage in voluntary activity and join certain kinds of political campaign. (Bentley and Oakley 1999: 13)

White, Bruce and Ritchie found that young people ranked education, employment, personal finance, discrimination, substance abuse, social and leisure facilities, and local issues among the issues that were most important to them (2000: 6–10). This suggests, of course, that young people are attracted to news about issues that are more likely to affect them.

Complicating matters still further, other researches indicate that young people are also interested in the same kinds of news topics as their elders. In the White et al. study, the young people surveyed expressed interest in crime and personal safety, taxation, health and international relations. Similarly, on the basis of surveys conducted with over 700 British youths, Henn and Weinstein concluded that 'this youth generation embraced a wide range of concerns':

> nearly half of our sample were preoccupied with mainstream concerns about public services (health 28 per cent, education 14 per cent, transport 4 per cent) and traditional 'materialist' issues (economic matters 7 per cent, Europe 6 per cent, crime and law and order 6 per cent). A noticeable minority (13 per cent) prioritised 'postmaterialist' concerns about wars and militarism, solidarity with the third world, animal rights and environmental protection. (Henn and Weinstein 2006: 522)

It is problematic, therefore, to think of young people's interests and concerns as alien to those of older generations. Young people as a group may be more interested in tabloid issues than their elders, and more drawn to news about issues such as education, employment and discrimination, which have greater relevance to their own lives. But it is a mistake to think that these topics, along with a handful of 'single issue' causes, are all they will pay attention to.

What these findings highlight, then, is that it is not just *what* is reported in the news that is important, but *how*. We noted earlier that many studies have concluded that young people are interested in political *issues* yet often find the *process* of formal politics repellent. This implies that there is potential to engage younger citizens with politics and public affairs, if only it were packaged differently. Similarly, it is not necessarily the case that young people find it more difficult to engage with 'hard' news *topics*. It may rather be factors such as the stuffy formality of the presentation, or an overwhelming focus on officials and socio-political 'elites', that makes 'hard' news less compelling for many younger viewers. Against such modes of television news, 'soft news' may well be an attractive alternative.

News diversity and political inclusion

Feelings of exclusion and marginalisation from politics and public affairs are a major cause of young people's lack of interest and participation in these areas. This conclusion appears repeatedly in the research literature (e.g. Bentley and Oakley 1999; Hall and Williamson 1999; Henn and Weinstein 2006; Power Report 2006; White, Bruce and Ritchie 2000). For Buckingham, young people's 'sense of *exclusion*' from politics and 'dominant forms of political discourse' arise from the fact that: 'Policies that directly concern young people – in areas such as education, family welfare, and leisure provision – are generally devised with little attempt to consult them or to gather their views' and thus politics is by and large 'conducted over their heads' (Buckingham 2000a: 251). Accordingly

> young people's lack of interest in politics could be seen as merely a rational response to their own powerlessness. Why should they bother to learn about something when they have no power to influence it, and when it makes no effort to address itself to them? (Ibid.)

Buckingham's point has clear implications for newsmakers' approach to younger viewers. By better incorporating young people's interests, concerns, viewpoints and tastes – and indeed more young persons themselves – into news provision, newsmakers could theoretically help bridge the gap between politics and news on one side, and young people on the other. As Buckingham continues

> Much greater efforts need to be made, not merely to explain the causes and the context of news events, but also to enable viewers to perceive their relevance to their everyday lives. News can no longer afford to confine itself to the words and actions of the powerful, or to the narrow and exclusive discourses that currently dominate social and political debate. (Ibid.: 253)

Similarly, White et al. concluded that one key to combating young people's political alienation was for politics and public affairs 'to be delivered in a more enjoyable and entertaining way than at present' and 'framed in terms which resonate with the issues and concerns of young people' (White, Bruce and Ritchie 2000: 40). Newsmakers also need to be better at 'involving young people in the programme, either on location or in the studio, on panels or through interviews' (ibid.: 41).

Lewis, Inthorn and Wahl-Jorgensen echo this conclusion. They argue that newsmakers contribute to political inactivity by presenting politics in a way that largely excludes 'ordinary people' and relegates the average citizen to the position of spectator. They describe the news as too often seeming 'like a soap opera for the chattering classes', while 'for those outside this circle – the more intermittent and less avid audience – news only serves to increase a sense of distance and insignificance' (Lewis, Inthorn and Wahl-Jorgensen 2005: 7).

This study also found that when the public did contribute, only a small minority of the ordinary people's quotations were 'deliberative' to the extent that they critically engaged with policy or made proposals. Much of their speech was about sport, consumer issues or personal experiences. When they were shown to be engaging with political developments, it was usually to react to politicians' statements and actions rather than critically discuss policy alternatives. Their comments also tended to be short sound bites and 'vox pops' (ibid.: 42–9).

Lewis, Inthorn and Wahl-Jorgensen's central thesis is that traditional notions of active citizenship are being replaced by a notion of the citizen as a kind of passive consumer of politics. 'Citizens are actively engaged in the shaping of society and the making of history; consumers simply choose between the products on display' (ibid.: 5–6). The net effect is that politics and public affairs become matters in which ordinary people are apparently not supposed to participate, outside of performing various basic actions such as observing what goes on and casting a vote at elections.

> We believe that it is not surprising that citizens feel alienated from politics given the limited and passive role they are allowed to play...In this way, the political and media establishments produce the alienation they claim to deplore. (Brookes, Lewis and Wahl-Jorgensen 2004: 78)

Lewis, Inthorn and Wahl-Jorgensen's work relates to the British public at large but, as they observed, political exclusion and political inactivity were most pronounced among the younger generation (2005: 56).

Supporting this hypothesis is decades of research literature on how news media report extra-parliamentary citizen activity, such as industrial strikes and protest rallies. When we consider instances such as strike action or protests against wars, laboratory testing on animals, or free-market globalisation, to name a few, news coverage tends to frame the action in terms of disturbance, disruption, violence and damage. As Gitlin (1980) outlined in his classic account of student protests from the mid-1960s, demonstrations are often reported through a dominant 'law and order' news frame that implicitly criminalises those taking part. Moreover, the message the protesters wish to convey is often lost as attention is fixed on the 'spectacle' of the event itself, or because opponents of the action – usually drawn from a social and political 'elite' – receive significantly greater opportunity to interpret and explain the issue at stake (Gitlin 1980; McLeod and Detenber 1999). By undermining the legitimacy of citizen activism when it transcends certain narrow boundaries, news media may thus reinforce the idea that politics is something best left to the professionals.

With respect to protests involving young people specifically, media analysis has shown that the coverage is often negative and condescending. Reporting on anti-globalisation protests, such as the demonstrations against the World Trade Organization summit in Seattle in 1999, is, for example, often marked by 'continual emphasis on violence, anarchists and property destruction' while the job of defining and steering debate around the central issue of free-market globalisation has mostly been given to establishment figures (e.g. McFarlane and Hay 2003). Young people were also prominent during the anti-Iraq war protests of 2003. The coverage began fairly sympathetically (before the war, when anti-war sentiment was considered more 'mainstream'), but when war became inevitable the thousands of young protesters were characterised as rowdy truants and their views were largely absented or trivialised (Cushion 2007; Such, Walker and Walker 2005). Given the concerns about young people's supposed apathy, their involvement in such campaigns could have been cause for optimism. Instead, media coverage has focused mainly on 'negative' elements that may undermine the legitimacy of the action. For newsmakers to report youth activism in these terms hardly sends a message to young people that their participation in politics is either welcome or taken seriously. Indeed, surveys of young people by Cushion revealed that 'news coverage of the anti-war protests discouraged some young people from participating in the political public sphere' (2009: 123).

One key to reversing trends towards disengagement and political alienation seems, therefore, to be a shift to more inclusive kinds of news-making that are less dominated by 'elite' actors and more open to 'ordinary' people. With respect to the younger generation, this means more coverage of young people and issues that involve, affect or concern them. It also means more opportunities for young people to contribute directly to news discourse, and implies the need for more positive and less condescending representations of political activism by young people and the public in general. Such measures can potentially erode the perception of politics as something alien, inaccessible and even irrelevant.

Research design

To investigate the impact of media content on public attitudes to politics and citizenship we conducted a detailed content analysis of British television news programmes. The analysis covered nightly bulletins from each of the five terrestrial channels (BBC1 22.00, BBC2 *Newsnight* 22.30, ITV1 22.30, Channel 4 19.00 and Five 17.30), as well as Sky News' nightly *Sky News at Ten* (22.00) and BBC1's youth news programme *Newsround* (17.25). This produced a sample of 2304 stories from 197 bulletins and 111 hours of news programming. All 197 bulletins were coded in full and results were added to a computerised database (Microsoft Access). In addition to this news sample we conducted an analysis of leading current affairs programmes (discussed in Chapters 6 and 7).

We analysed every episode of each of these programmes for one calendar month in 2006 (1 – 31 May 2006). We chose an intensive one-month analysis over a more staggered coding period so as to better witness the organic development of stories over time. News agendas may, of course, differ from one month to the next, and we reflect on this point where appropriate. Our selection of May 2006 as a coding month was semi-random in that the only relevant considerations were (a) that, in line with our predetermined research schedule, it had to be in the first half of 2006, and (b) that the sample would reflect as far as possible a 'normal' news month (i.e. outside of holiday and slow-news periods, when Parliament was in session, and before the summer 'silly season').

This chapter presents overall findings for (1) the topics or 'story subjects' reported (2) the individuals, groups or 'story actors' present, and (3) the 'news sources' quoted, in news stories. By analysing these three areas we are able to investigate two key areas of particular interest. First, via analysis of 'story subjects' we can map the contours of British television news agenda, revealing which subjects and themes are reported most frequently. Second, via analysis of 'story actors' and 'news sources' we can reveal which social groups receive the most attention and the best access to the news. Analysing both indices of news prominence also allows us to identify instances of groups who may receive much attention but are given comparatively little opportunity to define how they and their actions are framed.

In conducting our content survey, our main unit of analysis was the *news story*. Most of the results that follow are based on the frequency of different *story subjects* and *story actors* in terms of the total number of stories in which they appeared. Where indicated we also provide results based on story time.

Story subjects

When recording story subjects, each story was analysed for the presence of different subjects drawn from a detailed list of subject variables. This list was comprehensive, covering a wide range of social issues (e.g. environment, abortion, immigration), along with various sub-topics for broad subjects such as crime (e.g. sex offences, violent crimes), politics (e.g. 'competence/integrity', 'interparty conflict') and young people (e.g. 'high achievement/creditable behaviour', 'young celebrities', 'criminality', 'victims/vulnerability'). This enabled us to explore, quantitatively, key areas of importance to our research objectives in considerable depth. Each of these subject variables fell into one of 25 broader subject categories such as 'Politics', 'International News', 'Crime', 'Health', 'Immigration' and 'Sport'. The analysis below is based on the results for these 25 broad categories. In subsequent chapters smaller subcategories receive fuller attention as we look more closely at issues such as politics, young people, and law and order. At certain points, the 25 subject categories have been condensed further into 8 broader categories which have been adapted from earlier research (e.g. Hargreaves and Thomas 2002).

Subject prominence

Because some subjects are more prominent in news stories than others, we coded each story subject with a prominence ranking from *'principal'* ('1') to *'main'* ('M') to *'subsidiary'* ('S'). On the first level, we coded each story with a single *'principal'* topic which best described what the story was about. This often proved difficult as many stories had two subjects which seemed equally appropriate, for example, 'Politics' and 'Terrorism' in stories about the release of two parliamentary reports into the 7 July 2005 terrorist attacks in London. Coders were, however, able to agree on the *'principal'* topic for 89 of 100 stories that were randomly selected for intercoder reliability testing. This method of coding, using a single-coding methodology, is consistent with most other news agenda analyses and has the advantage of providing 'clean' statistics (adding up to 100 per cent). However, because news stories will not always limit their focus conveniently to one or another pre-selected topic, this method will inevitably under-record the level of attention given to some topics. Certain subjects, such as 'Crime' and 'Politics', will often appear in connection with a second topic, such as in the politics/terrorism example given above.

In order to achieve a broader overview of news subjects, we therefore coded all other subjects which received more than passing notice, giving them a ranking of *'main'* or *'subsidiary'*. *'Main'* subjects were those which, like the *'principal'* topic, were very prominent in the story. *'Subsidiary'* subjects were any topics which received significant attention but were clearly subsidiary to the central story themes. Intercoder reliability across the three levels of analysis (1/M/S) was 80 per cent.[1] In the statistical results below we provide results for these three different levels of subject prominence but focus most attention on the combined results for *'principal'* and *'main'* subjects (*'principal/main'* or '1/M').

Story actors

In addition to story subjects we recorded the presence and prominence of different story 'actors'. Through this measure we recorded all the persons, organisations and groups that appeared in each story. We recorded both those who had the opportunity to speak and those who were spoken about. Our variable list for story actors was also comprehensive, allowing, for example, for divisions across party lines for 'politicians', and according to gender and ethnicity for 'young people'. In this chapter, we focus primarily on results for 16 broader categories of story actors. As with our analysis of story subjects above, actors were coded with a prominence ranking of *'main'* or *'subsidiary'* (there was no coding of *'principal'* story actors). Intercoder reliability for *'main/subsidiary'* (M/S) story actors was 94 per cent across broad categories (see note 1).

News sources

In addition to recording how frequently various types of story actor appeared on the news, we measured their speaking opportunities as 'news sources'. These speaking opportunities consisted primarily of interviews, talking head sound bites and person-on-the-street 'vox pops'. (Speech by news anchors and reporters was excluded.) The results for news sources have been analysed according to both total *speaking time* and the number of individual *speaking opportunities*. When coding for news sources, we recorded who spoke, how long they spoke for, and the topic they were speaking about. Intercoder reliability was 91 per cent for identifying which story actor was quoted, 98 per cent for quotation length and 91 per cent for the topic of speech.

Results

'Principal' story topics

In line with the quantitative analyses cited above, we found that serious topics tended to be more prevalent than 'tabloid' themes on television news bulletins. Table 4.1 outlines results for *'principal'* subjects, arranged into 8 broad categories. (Results are provided both for the total number of stories that focused on these topics, and for the proportion of total story time they received.) By either measure, the most frequent news topics were international news, social and economic issues (taken collectively), and British politics. The 'tabloid' issues of sport, crime and entertainment, lifestyle and 'human interest' news were the principal focus in fewer stories. News items dealing with these topics also tended to be shorter, as reflected by the statistical variations for story numbers and story time. While 14.6 per cent of stories were primarily about sport and 14.1 per cent about crime, these stories comprised only 8.9 per cent and 10.4 per cent of

Table 4.1 Principal story subjects

Subject	Stories	Per cent	Minutes	Per cent
International News	557	24.2	1,167.2	25.9
Social/Economic	418	18.1	932.4	20.7
British Politics	382	16.6	963.4	21.4
Sport	336	14.6	401.5	8.9
Crime	325	14.1	467.6	10.4
Human Interest	155	6.7	229.4	5.1
Terrorism	47	2.0	140.7	3.1
Other	84	3.6	197.5	4.4
Totals	2,304	100.0	4,499.4	100.0

total story time respectively. Entertainment, lifestyle and 'human interest' items also tended to be shorter than average. In contrast, percentages for story time were higher than those for story numbers for all four of the 'hard' news topics (including terrorism). Taken together, stories about 'hard' topics lasted 2 minutes 17 seconds on average, while those about 'tabloid' topics lasted 1 minute 21 seconds. 'Hard' news stories were thus almost a minute longer on average than 'tabloid'.

Stories dealing with hard news topics were, therefore, both more frequent and longer on average than those concerned with tabloid themes. Altogether, the 'tabloid' topics comprised 35 per cent of all stories (816 of 2304 stories) and 24 per cent of the total story time (1099 of 4499 min).[2]

Looking now at results for each of the individual programmes in our sample, we found that BBC2's *Newsnight* had the most serious news agenda. Table 4.2 outlines the results for *'principal'* story subjects for each programme based on story numbers (rather than time): 37.3 per cent of *Newsnight* stories were principally about domestic politics, 22.3 per cent international news and 20.0 per cent social and economic issues. Just 5.5 per cent of stories were primarily about crime, 2.3 per cent sport and 1.8 per cent entertainment, lifestyle and 'human interest' items. At the other extreme was Channel Five's *Five News*. Five had the lowest proportion of political news stories (11.2 per cent) excluding *Newsround*, and the lowest level of international news for all programmes (12.4 per cent). It did have one of the highest rates of news about social and economic issues (mainly social), at 22.4 per cent. Otherwise, the *Five News* agenda was marked by the prominence of crime (22.9 per cent), sport (19.1 per cent) and entertainment, lifestyle and 'human interest' news (9.1 per cent).

Table 4.2 Principal topics by news programme (percentages only)

Topic	BBC	NN	ITV	C4	C5	Sky	NR
International	27.6	22.3	21.0	32.3	12.4	26.5	18.0
Social-economic	17.5	20.0	13.9	18.7	22.4	15.0	22.7
British politics	17.8	37.3	17.0	14.9	11.2	16.3	1.3
Terrorism	2.2	1.8	2.5	2.5	1.5	2.1	0.7
Crime	12.3	5.5	13.9	12.8	22.9	16.8	9.3
Sport	15.0	2.3	24.4	10.5	19.1	13.4	17.3
Human interest	4.1	1.8	5.6	4.6	9.1	7.6	22.7
Other	3.6	9.1	1.9	3.6	1.5	2.4	8.0
Totals	100.0	100.0	100.0	100.0	100.0	100.0	100.0

Newsround is a programme directed at children and not surprisingly its news focus was rather unique. Only 1.3 per cent of *Newsround* stories focused primarily on politics, while the soft news themes such as sport (17.3 per cent) and entertainment, lifestyle and 'human interest' (22.7 per cent) were among its most prevalent subjects. However, *Newsround* had the highest proportion of stories about social and economic issues (22.7 per cent), particularly the environment (10.0 per cent) and education (4.7 per cent). It also had a reasonably high level of international news (18.0 per cent) and one of the lowest proportions of crime news (9.3 per cent).

There were also notable differences between the remaining four programmes, yet these programmes shared various similarities in news focus as well (when compared with *Newsnight*, *Five News* and *Newsround*). Of these four, *Channel 4 News* and BBC1's *Ten O'Clock News* devoted most attention to 'hard' news topics. Altogether, 68.4 per cent of Channel 4's stories dealt primarily with politics, international news, social and economic issues, and terrorism. For the BBC this figure was 65.1 per cent. Sky and ITV also paid greater attention to the 'serious' issues, yet had lower aggregate levels of hard news overall – Sky with 59.9 per cent and ITV with 54.4 per cent. The figure for *Newsnight* was 81.4 per cent, for *Five News* 47.5 per cent and for *Newsround* 42.7 per cent. When the same is measured according to the amount of *time* given to each topic, the percentage of 'hard' news content increases markedly for every programme (*Newsnight* – 95.3 per cent, *Channel 4 News* – 82.2 per cent, *BBC News* – 78.3 per cent, *Sky News* – 65.8 per cent, *ITV News* – 64.3 per cent, *Newsround* – 56.8 per cent, *Five News* – 54.1 per cent).

'Main' and 'subsidiary' subjects

To achieve a more comprehensive overview of how often each subject appeared on the news, we have also presented results for *'main'* and *'subsidiary'* story subjects. Table 4.3 outlines the total number of stories in which different news subjects were coded *'principal'* or *'main'* (1/M), and thereafter *'principal'*, *'main'* or *'subsidiary'* (1/M/S). The results presented here are more detailed than those given above, with for instance the broad category of 'social and economic issues' divided into individual topics.

As is to be expected, the addition of *'main'* and *'subsidiary'* subjects produces higher percentages across the board. The most dramatic increases are for the subjects 'politics' and 'crime'. Politics was often a main theme in stories about social and economic issues and was present

Table 4.3 'Principal/main/subsidiary' story subjects

Topic	Principal/Main	Per cent (n = 2,304)	1/M/S	Per cent (n = 2,304)
International news	649	28.2	696	30.2
British politics	517	22.4	582	25.3
Crime/Law and Order	472	20.5	529	23.0
Sport	361	15.7	370	16.1
Human interest	233	10.1	269	11.7
Economic/Business	196	8.5	236	10.2
Health/Medicine	163	7.1	186	8.1
Immigration	102	4.4	155	6.7
Environment	69	3.0	98	4.3
Media	65	2.8	87	3.8
Terrorism	59	2.6	65	2.8
Education	53	2.3	62	2.7
Accident/Disaster	51	2.1	61	2.6
Animal rights	38	1.6	54	2.3
Religion/Ethnicity	38	1.6	39	1.7
Science/Technology	28	1.2	37	1.6
Employment	28	1.2	30	1.3
Transport	25	1.1	37	1.6
Gender/Sexuality	25	1.0	33	1.4
Arts/Culture	22	1.0	24	1.0
Human rights	22	1.0	26	1.1
Energy	17	0.7	19	0.8
Gambling	5	0.2	5	0.2
Abortion	3	0.1	3	0.1
Housing	2	0.1	2	0.1

in a large proportion of the terrorism stories (as in the example given above). Similarly, crime and law and order issues were a main or subsidiary theme in a large number of politics stories. In particular there were many items relating to a political scandal over the Home Office's handling of foreign criminals.

Limiting our discussion to results for *'principal'* and *'main'* subjects, we found that international news remains the most frequent topic, present in 28.2 per cent of stories. Thereafter come British politics (22.4 per cent), crime (20.5 per cent) and sport (15.7 per cent). One in ten stories (10.1 per cent) were judged to contain a significant entertainment, lifestyle or 'human interest' element. Examples of stories with a strong 'human interest' focus were a yacht sinking after the owner let his wife steer (2 May), Tony Blair wearing a red wristband during Prime Ministers Questions (3 May), Conservative politician Boris Johnson making a dangerous tackle during a charity soccer match (4 May), an obese man walking across America to lose weight (9 May), a taxi driver mistakenly interviewed as an IT expert on BBC 24 (15 May), and a British expedition's attempt to climb Mt Everest (recurring item on Sky and Five).

These results reveal an increase in the number of stories in which 'tabloid' subjects were prominent, yet overall they confirm the above conclusion that 'hard' news topics were more prevalent than 'soft'. Hard topics remain the most frequent overall and, as noted above, tabloid topics were the most likely to be reported in brief. After human interest stories, all other story subjects fell below 10 per cent of television news coverage. Economic issues constituted 8.5 per cent of coverage and health/medicine 7.1 per cent. After that all other issues fell below 5 per cent of the output such as the environment (3.0 per cent), education (2.3 per cent), religion and ethnicity (1.6 per cent), and animal rights (1.6 per cent).

Story 'actors'

When investigating the prominence of different individuals and social groups in news stories, we were especially interested in identifying the frequency with which the public – or so-called ordinary people – appeared on the news. We found that the public were prominent as *'main'* actors in more than one in three stories (36.7 per cent), and *'main'* or *'subsidiary'* in over half (52.8 per cent). This is much higher even than for politicians, who were the second most prominent group and were *'main'* in 22.1 per cent of stories, and *'main'* or *'subsidiary'* in 28.4 per cent (see Table 4.4). The most prominent types

Table 4.4 Story actors

Story Actors	**Main**	**% (n = 2,304)**	**Main/Sub**	**% (n = 2,304)**
Ordinary people/ The public	845	36.7	1,217	52.8
British politicians	509	22.1	654	28.4
Police/Military	341	14.8	502	21.8
Celebrities	270	11.7	303	13.2
International leaders	209	9.1	324	14.1
Business	146	6.3	218	9.5
'Experts'	140	6.1	416	18.1
Public servants/ Govt dept.	126	5.5	235	10.2
'Terrorists'	63	2.7	88	3.8
Religious groups/Leaders	30	1.3	48	2.1
NGOs/Pressure groups	28	1.2	114	4.9
Judges/Judiciary	27	1.2	88	3.8
Royal family	22	1.0	23	1.0
Teachers/Education	21	0.9	54	2.3
Trade unions	18	0.8	41	1.8
Community workers	0	0.0	8	0.3
Other	26	1.1	83	3.6

of story actor after the public and politicians were police, security and the military personnel (*'main'* in 14.8 per cent of stories), 'celebrities' (11.7 per cent), international leaders (9.1 per cent), businesspersons (6.3 per cent), 'experts' of various descriptions (6.1 per cent), and public servants and officials (5.5 per cent).

When recording appearances by the public, we recorded instances when individual 'ordinary people' either spoke or were spoken about, as well as references to the public in its various guises (e.g. 'taxpayers', 'voters'). Instances of the latter were not just passing remarks, as there were many items in which 'the public' – as an abstraction – was a major focal point. These included stories about immigrants, foreign prisoners, and voters in the lead-up to local government elections on 4 May. Invocations of the public as abstractions ought, however, to be seen as something of a 'phantom' story actor, one in which the public is frequently recruited as it were to support the argument or perspective of another who claims to speak on behalf of some unified collective body.

With respect to focus on individual (actual, concrete) 'ordinary people', we kept our notion of who belonged to this category quite broad. In general, this included anyone who was *not* included in the story on the basis of any formal expertise or qualifications; was *not* identified as representing a group or organisation; and did *not* hold any official status nor belong to a social, political, cultural or economic 'elite' (e.g. politicians, police, military, judges, celebrities, royalty, businesspersons). This produced a category which spanned diverse social groups including non-managerial workers, students, protesters and activists, 'civilians' in conflict-torn regions, immigrants and asylum seekers, victims of crime (including their families), criminals and crime suspects, and of course the 'person-on-the-street' from 'vox-pop' interviews. Given the breadth of the category, it is less surprising that 'the public' was present in so many stories.

News sources

Ordinary people were also frequently quoted in our sample. No other category of social actor was in fact quoted as frequently. As outlined in Table 4.5, 32.1 per cent of all sources for quotations were ordinary people. This is significantly higher than any other group, including politicians (20.6 per cent) and 'experts' (15.2 per cent). However, if we measure the public's share of quotations by the amount of speaking time they received, they come in third behind politicians, who received 29 per cent of speaking time, and 'experts' (20.2 per cent). Altogether the public contributed 19.1 per cent of speaking time by news sources.

Ordinary people tended to receive relatively short speaking opportunities, however. Each individual ordinary person news source spoke for an average of 14 seconds (per story, not per individual sound bite). The average speaking time for politicians was 34 seconds, 33 seconds for international leaders, 32 seconds for 'experts', 27 seconds for public servants and 24 seconds for businesspersons. (Our category of 'experts' included academics, people from think tanks, journalists, writers, artists, 'intellectuals' and scientists, as well as doctors, lawyers and other professionals who were implied to possess relevant expertise to the issue at hand.)

After politicians, experts and the public, the story actors who received most speaking time were international leaders (7.3 per cent), 'celebrities' (5.1 per cent), businesspersons (4.6 per cent), and police, security and military personnel (4.0 per cent). With respect to diversity of viewpoints, it is worth mentioning that social actors who might be expected to hold viewpoints which conflict with the dominant political ideas at Westminster, such as NGOs and pressure groups (3.3 per cent), religious

Table 4.5 News sources

News sources	Seconds	% (n = 79,716)	Sources	% (n = 3,313)
British politicians	23,322	29.3	681	20.6
'Experts'	16,102	20.2	503	15.2
Ordinary people/ The Public	15,246	19.1	1,065	32.1
International leaders	5,784	7.3	177	5.3
Celebrities	4,034	5.1	192	5.8
Business	3,672	4.6	150	4.5
Police/Military	3,198	4.0	197	5.9
NGOs/Pressure groups	2,664	3.3	102	3.1
Public servants/ Govt dept.	1,970	2.5	73	2.2
Judges/Judiciary	871	1.1	25	1.1
Teachers/Education	762	1.0	37	1.0
Religious groups/Leaders	665	0.8	30	0.8
Trade unions	613	0.8	27	0.8
Royal family	126	0.2	5	0.2
'Terrorists'	113	0.1	11	0.1
Other	574	0.7	38	1.1
Totals	79,716	100.0	3,313	100.0

groups (0.8 per cent) and trade unions (0.8 per cent), received comparatively little speaking time.

Looking at the results for each programme separately (see Table 4.6), we find that the programmes with a 'softer' news agenda tended to rely less heavily on 'elite' sources such as politicians, international leaders and 'experts'. Given that politicians tend to dominate within political stories and that ordinary people are most prominent in news about crime and social issues (discussed below), the connection between soft news and non-elite sources is not unexpected. The two extremes were again represented by *Newsnight* on one side, where politicians contributed 41.0 per cent of quotes and the public only 5.5 per cent, and *Five News* on the other where politicians contributed 11.5 per cent and the public 42.7 per cent. *Newsround* was again clearly removed from the other programmes, with ordinary people receiving 70.6 per cent of speaking time compared with 0.6 per cent for politicians. Regular people were in fact the most quoted category of story actors on BBC, ITV and Sky as well, in each case receiving approximately one-quarter of all speaking time.[3]

Table 4.6 News sources by programme (percentages only)

	BBC	NN	ITV	C4	Five	Sky	NR
Politicians	24.1	41.0	22.5	33.2	11.5	21.2	0.6
'Experts'	16.4	28.1	14.9	17.3	18.2	17.1	13.8
Ordinary people/ Public	26.3	5.5	24.4	12.8	42.7	26.5	70.6
International leaders	5.8	7.3	4.2	12.1	0.8	5.4	0.0
Celebrities	6.0	2.9	13.5	3.1	7.5	6.6	8.6
Business	4.3	5.4	3.2	5.8	2.0	3.7	1.7
Police/Military	6.3	0.5	8.0	4.3	6.5	6.9	0.0
NGOs/Pressure groups	3.3	2.6	2.8	4.6	4.0	2.4	1.2
Other	7.6	6.7	6.6	6.9	6.9	10.0	3.5
Totals	100.0	100.0	100.0	100.0	100.0	100.0	100.0

Discussion

As outlined above, efforts to sketch out a young person's news agenda have produced conflicting results. While research into young people's news consumption often shows them to be disproportionately attracted to 'soft news', other surveys and focus groups reveal significant interest in a broad range of serious issues. Addressing first the question of 'soft news', we found in line with past studies that television news contains significant levels of tabloid content. Only one programme, BBC2's *Newsnight*, broadcast very low levels of soft news. On Channel Five's *Five News* and on the BBC1 children's news programme, *Newsround*, soft news items comprised approximately 50 per cent of all stories. If we assume then that young viewers are disproportionately interested in 'soft' news topics, they would seem to be fairly well catered for by several of Britain's television news programmes. This is in terms of the narrow question of a preference for serious or light news, at least.

The varying levels of hard and soft news between the seven programmes highlights that there is also a fair degree of choice available for British television news audiences – again at least in terms of the broad brush categories of 'hard' versus 'soft' content. At one end, *Newsnight* might be expected to appeal to citizens who are already politically engaged and who are mostly interested in hard news. At the other, *Five News* offers a news service that is lighter on politics and other 'hard' themes while relatively heavy on tabloid content. Moreover, *Newsround*, by design, attempts to provide a news service that is accessible and appealing to younger audiences, especially children. As to the question of whether there is too much soft news on television, our results confirmed those of previous studies that, on most programmes at least, hard news topics continue to receive the most attention. Across most channels hard news stories were more numerous than tabloid items, and were significantly longer on average. With this in mind it seems that our analysis has confirmed Barnett et al.'s conclusion that British television news provides a balance of serious and light news.

Moving on from the question of soft news, research cited above suggests that young people are easily engaged with hard news themes as well, whether 'single issue' ('post-materialist') causes or issues related to their own lives and experiences. Specific topics that are often named in the research literature include wars, poverty, human rights, animal rights, discrimination, the environment, education, employment and social/leisure centres. Here we start to see in what ways broadcast television news is failing to construct a television news agenda more consonant with the interests and concerns of young people.

It is a notable result that most of the above-cited issues did not receive much attention in our news sample. *Newsround* was the only real exception here, devoting considerable attention to education and the environment. Measured by the time given to these topics, they comprised 6.6 per cent and 16.6 per cent of *Newsround* content respectively. Returning to the aggregated results, even when we include the occasions in which these topics were '*subsidiary*' story emphases, their share of the overall news agenda rarely comprises more than a couple of percentage points. Across the full sample, 2.7 per cent of stories dealt with education and 1.3 per cent with employment. Only a handful of stories dealt with social and leisure centres, which tended to appear in stories concerned with the problem of juvenile crime. 'Post-materialist' issues received only marginally greater attention: 4.3 per cent of stories dealt with the environment, 2.3 per cent with animal rights and 1.1 per cent with human rights.[4] Similarly, although there were many international news stories in our sample, efforts to alleviate poverty and promote global justice were rarely addressed. Our analysis suggests that the distribution of coverage across story topics is rather weighted towards a small number of traditional, 'mainstream' issues such as politics, crime, health and the economy. As noted earlier, some research has found young people to be interested in these matters of course, but what is missing is a more even distribution across a larger range of topics that might well open the news up to a wider audience. The obvious problem with the relatively scant attention given to issues of concern for young people is that it could reinforce the idea of news, and possibly politics and public affairs more broadly, as something that is 'out of touch' with their lives and interests. If, for example, political coverage as it is currently done is a turn-off for younger viewers, then a quarter of the output constitutes a disincentive to view. If international news, as it is currently done, is also too remote and distant for some young people (and the content of the tabloid press which is read overwhelmingly by the working class would seem to indicate that 'international news' may be too 'abstract' for this socio-economic group) then combined with the politics stories, over half of the coverage is a disincentive to view. In other words, a more appealing news agenda might be one which would have a different and more equitable spread of coverage across the story subjects than is evident in our sample.

With respect to broadcasters' effectiveness in including young people, and ordinary people more generally, we found mixed results. We explore how broadcasters represented young people specifically in Chapter 5, but in terms of the access given to ordinary people generally we found that the public received, in overall quantitative terms, significant attention in over half of all stories, appearing even more frequently than politicians. Ordinary people also received a greater number of *individual* quotations than all other categories of story actor, comprising one-third of quoted sources (32.1 per cent).

However, upon closer inspection these apparently promising results become a little more problematic. For instance, if we analyse news quotations according to the amount of time allocated to different groups of news sources, we find that the public ranks third behind politicians and 'experts'. Moreover, when it comes to the quality of ordinary people's speaking opportunities overall, we found that members of the general public spoke for significantly shorter time intervals than politicians, 'experts' and other 'elite' actors.[5] Shorter quotations usually mean diminished opportunities for being able to express sophisticated opinions and develop meaningful arguments.

Another somewhat problematic finding is that ordinary people appeared most often in connection with crime news. In 41 per cent of stories where the public was a *'main'* story actor, crime was a *'principal'* or *'main'* subject (n = 344 stories). Similarly, when ordinary people spoke as news sources in interviews, sound bites and 'vox pops', 23 per cent of their speaking time related to this topic. A considerable proportion of prominent 'ordinary people' in the news were thus victims and perpetrators of crime, or their families and friends, and hence much of their speech was emotional or articulated personal experiences only, or views that were grounded in a specific incident. Generally, ordinary people are not asked to comment on issues of policy or articulate broader political points.

As we discuss in Chapter 6, the public's opportunities for contributing to news discourse about 'politics' were rather limited. Ordinary people received only 9 per cent of speaking time by news sources in stories about politics. This is compared with 62 per cent for politicians and 16 per cent for 'experts'.

Conclusion

In this chapter we have considered whether the topics covered on television news are likely to appeal to young viewers, and how effective broadcasters are at meeting the public's demand for news and views of 'ordinary' people just like them. With respect to whether broadcasters deliver a service that young people will want to watch, the results are mixed. On the positive side, news programmes cover a reasonable range of topics and there is significant variation in the topic agendas of the programmes available on terrestrial television. On the negative side, many of the topics that are supposed to be most important to young people received relatively little attention. This is true both for the 'post-materialist' themes and issues with direct relevance to young people's lives.

With respect to the role ordinary people play in television news, we found mixed results as well. Broadcasters mention or even focus on the public quite a lot; most programmes allow ordinary people significant opportunities to contribute to news discourses as quoted sources. In most programmes ordinary people received both a greater number of individual quotations and more overall speaking time than other categories of story actor. Across the full sample they were third with respect to total speaking time, though on five of the seven programmes they received more speaking time as well. However, ordinary people received considerably shorter speaking opportunities compared with 'elite' actors on average, and this was consistent across most channels.

Numerous studies, including the Lewis, Inthorn and Wahl-Jorgensen's *Citizens and Consumers* (2005) and the BBC's own *Beyond the Soundbite*, have pointed to the need for greater inclusion of the public and our results suggest that the broadcasters are trying to meet these recommendations. However, this inclusiveness could be broadened to encompass news about politics and economic and industrial news, among other hard news issues. Currently ordinary people are disproportionately visible in relation to tabloid topics, especially crime. In news about social issues they were often present, though not always in the role of the concerned and engaged citizen. Two of the most reported health-related news stories involving ordinary people were, for example, news that a 62-year-old woman had used IVF treatment to become pregnant, and that an old-age pensioner had travelled to France to have surgery because the domestic waiting list was too long. From a qualitative perspective therefore, there remains room for improvement.

As far as the main news bulletins are concerned, they cannot, of course, be expected to cater exclusively for the youth market. There is perhaps room for one or more *individual programmes* aimed specifically at young people, just as *Newsround* is aimed at children. But what the main bulletins can do is to focus more attention on issues that interest and engage young people, and focus more on young people themselves. How much and how well television news portrays young people currently is the subject of the next chapter.

5
The Symbolic Criminalisation of Young people

In 2005, the Bluewater shopping mall in Kent UK banned young men from wearing 'hoodies' and caps (although not shops from selling them), articles of clothing that the media and politicians had increasingly associated with crime and violence (Waterhouse 2005: 16). The construction of young people as a 'problem' is certainly not new. As Osgerby (1997) and Pearson (1983) have shown, the demonisation of young people in the UK dates back at least to the eighteenth century and tends to rise and fall in accord with broader structural changes (and tensions) in society. In the post-Second World War era, within the context of a social democratic settlement and the establishment of a welfare state, young people were major beneficiaries of a new strategy of inclusion (Mizen 2004: 17). This post-war period certainly witnessed the creation of a veritable rogues gallery of 'delinquent' young people, from teddy boys, mods, rockers and skinheads through to punks, crusties and ravers. However, while particular subcultures of youth might be labelled within the media (and by other custodians of official morality) as 'deviant', they were also clearly seen as tiny minorities: precisely *subcultures*. Since the 1980s, with the dismantling of the welfare state, substantial deindustrialisation and the opening up of British society to global market forces, being a young person has become increasingly risky and the responsibilities for negotiating those risks have shifted away from collective provision and become highly individualised (Furlong and Cartmel 1997: 4).

At the same time, the sense of threat around media representations of young people has arguably become less and less restricted to subcultures of youth and increasingly prevalent across the category of youth itself. Perhaps for the first time since the post-Second World War period, youth itself has become a signifier for folk devils, figures who come to be 'defined as a threat to societal values and interests' (Cohen 2002: 1). While legislation by the then Conservative government enshrined a more punitive attitude towards young people in the 1982 Criminal Justice Act (Newburn 1996), the early 1990s saw a significant shift in the framing of young people in news media – notably the linkage of youth and crime within the British news media. The murder of two-year-old James Bulger by a pair of ten-year-old children in 1993 seems to have marked a defining moment in translating the emerging groundtone of state attitudes towards young people into popular culture representations as constructed within the news media (see Franklin and Petley 1996).

It has now become commonplace, as Hall and Williamson argue, for young people to feature 'prominently in the more lurid descriptions of an underclass', appearing commonly as, among other things, 'violent young criminals, thoughtless teenage mothers, or idle "dole-scroungers"' (1999: 11). It should be evident that the category of young people here strongly intersects with class: it is not the sons and daughters of the middle class sniffing cocaine in the back of taxi cabs who are being invoked as a threat in such representations, but the young working class. According to Griffin, the dominant images of young people in the media, and in society generally, are 'youth as trouble' and 'youth in trouble' (2004: 14). 'As trouble', youth may be in danger of coming to function as the source of a moral panic, where 'intense public hostility and condemnation' reinforce moves towards 'strengthening the social control apparatus of society' (Goode and Ben-Yehuda 1994: 31).

A number of commentators have noted that within the public sphere of Western highly market-orientated societies, such as the UK and the US, young people today are framed within a dualistic vision of being cutting edge consumers and fashion trend setters on the one hand and/or threats to public order on the other (Giroux 2003; Goodman 2003). To what extent does British television news provision reproduce this dualistic vision of young people? If television news 'enables viewers to construct and define their relationship with the public sphere' (Buckingham 2000a: 18), then as a normative ideal one would expect broadcasters to explore the complex mix of determinants shaping young people's lives and identities across a whole range of areas that might include education, jobs, housing, political representation and so forth.

The following analysis will suggest that television broadcasters are falling a long way short of this normative ideal. We will present and discuss a combination of quantitative and qualitative analysis of television news output to establish this argument. The statistical analysis will function to establish the frequency with which young people appear in the media and in relation to which stories. This statistical analysis provides evidence that our smaller qualitative case study has some more general validity in its discussion of media approaches to young people and crime. At the same time the qualitative analysis explores in more detail the audio-visual language of television news and teases out some of its contradictions. Together the quantitative and qualitative analyses seek to interrogate the ideological implications of television news. By ideological we mean the unquestioned assumptions, routine preferences and the systematic exercise of news choices that produce a representation of the world and the place of young people within it, that is uncritically congruent with social inequalities and requires limited investigation of dominant institutions (such as the state and the market) and dominant policy trends.

Methodology

We have already explained the coding methods employed for this project in the previous chapter. There are, however, a few points that we need to mention which relate specifically to how we analysed the representation of young people in our sample. The first of these is that this chapter's analysis is conducted over the same sample of 2304 news stories discussed in the previous chapter, yet it excludes 174 items that were simply straight reporting of sports results. In such stories there is usually no or little focus on identifiable individuals or social groups as opposed to scores and teams. We judged it problematic to include these stories in an analysis of young people, especially given that many sporting events will involve young persons (as athletes often are), even if this is not often clear in the reportage. We should also emphasise though, that we did include stories that related to sport but were not just straight reporting of results. Examples of such stories included the appointment of a new English football manager and the build-ups to the UEFA Champions League final, the FA Cup final and the FIFA World Cup. Subtracting these 174 sport result stories, our analysis of young people, therefore, covers a sample of 2130 stories.

Second, because we are looking specifically at the representation of 'young people', we analyse stories in which young people were identified as being a main focus (i.e. where the actions of young people, or events or issues involving young people, were central to the report). For coding purposes, and in order to incorporate a broad range of youth representations, we categorised 'young people' as teenagers and young adults from 14 to 25 years of age.

From our broader sample of 2130 stories we identified 286 stories in which young people were a main focus. Those 286 news stories were then coded in further detail. First we noted the frequency with which young people featured as 'story actors', and assigned each story actor with a prominence ranking (*'main'* or *'subsidiary'*). A common and recurring instance of a young person being coded *'main'* was reports focusing on 20-year-old football star Wayne Rooney's struggle to overcome a foot injury in time for the 2006 World Cup. Second, we measured young people's speaking opportunities as 'news sources', recording the length of their speech (in number of seconds) and noting the 'topics of speech' (as explained in the previous chapter). Third, we tested for the frequency with which various young people-specific 'story subjects' appeared in news stories, drawing from a list of 19 variables including 'young people as politically interested/active', 'young people and high achievement/good behaviour', 'young people as victims/vulnerable' and 'young people as violent criminals'. Each subject was given a prominence ranking to reflect the level of attention it received in the report (*'main'* or *'subsidiary'*). This detail allows us to map the profile of 'young people' in television news (the frequency of their appearances, the frequency of their use as news sources, and the kinds of actions and behaviours that receive news attention).

As with the previous chapter, the main unit of analysis is the news story. When we discuss the frequency of young people 'story actors', we are talking about the number of stories in which young people featured as *'main'* or *'subsidiary'* story actors – not the total number of all young people that featured on the news during our sample period. (That is, ten young people could appear in a single story, but this would be counted as only one story featuring young people).

The same applies for young people 'story subjects'. This measure records the number of news stories in which different 'story subjects' (topics, themes) involving young people (e.g. 'Crimes committed by young people', 'young people as celebrities') appeared during the sample period (as *'main'* or *'subsidiary'* emphases). Because these subjects were not mutually exclusive, it was possible to record multiple young people subjects for a single story (e.g. one story could have both 'Crimes committed by young people' and 'young people as victims/vulnerable'). This is reflected in some of the tables and statistics below where the total percentages exceed 100 per cent. (The percentages for each 'story subject' have been calculated by total number of stories, not total number of subjects.)

Most of the results below are for stories in which young people 'story actors' or different young people 'story subjects' were '*main*' foci. From this point forward therefore, when reference is made to 'young people stories', the reference is to stories in which 'young people' have been coded as being a '*main*' focus. The key exception to this 'story level' analysis is the results for young people as news sources, which are presented in total number of seconds.

Key results: Representations of young people in television news

As noted, we identified 286 stories in which events or issues involving young people were a '*main*' focus. As shown in Table 5.1, 47 per cent of these stories were about crimes committed by young people. Of these crime stories, almost three-quarters (72 per cent, n = 97) related to violent crime, with a further 18 per cent (n = 24) concerned terrorism. Ninety per cent of stories focusing on youth crime were, therefore, about violent crime or terrorism.

In more than half (n = 70) of youth crime stories, the victim was another young person. Altogether, there were 105 stories about crimes committed against young people, of which 89 per cent (n = 93) related to violent attacks. There were also 18 further reports in which young people featured heavily as 'victims' or 'vulnerable', meaning 43 per cent (n = 123) of stories about young people displayed them in some way or another as being 'at risk'. Crime accounted for 85 per cent of the 'threats', with the small remainder including a university lecturers' strike (n = 5), a disastrous drug trial experiment (n = 3), and inadequate funding for disability and mental health (n = 2).

Table 5.1 Frequency of young people as main subjects in news stories

Young people subjects	Stories	Per cent (n = 286)
Crimes committed by young people	134	46.9
Victims/Vulnerable	123	43.0
Young 'celebrities'	81	28.3
Achievement/Good citizens	73	25.5
Political interest	9	3.1
Teen pregnancy/Sex	3	1.1
Disadvantage/Unemployment	3	1.1
Other	13	4.5

Outside the world of 'crime', young people were most visible as 'celebrities'. Twenty-eight per cent (n = 81) of stories about young people centred on the actions of young celebrities, most notably footballers Wayne Rooney and Theo Walcott. With the FIFA World Cup only a month away, these two England strikers were the subject of 59 stories. Largely because of the overwhelmingly positive attention they, and other sporting celebrities, received, there was a fairly high number of stories concerning youth achievement (n = 73, or 26 per cent of young people stories). Outside of sporting and entertainment success, however, youth achievement received scant attention. Indeed, there were only 4 stories (1 per cent of young people stories) in our sample that focused on non-sport/celebrity achievements, or even good citizenly behaviour, by young people.

If we exclude the 'celebrity' stories from our statistical analysis and consider only stories in which 'non-celebrity' or 'ordinary' young people were a *'main'* focus (n = 205), the association with 'crime' becomes even greater. In 82 per cent of these stories (n = 169) young people are linked to crime either as victims or as perpetrators. Two-thirds (65 per cent, n = 134) related to crimes perpetrated by young people specifically. In the previous chapter we noted that 41 per cent of stories with all 'ordinary people' coded *'main'* were about crime. The figure for 'ordinary' young people specifically is therefore twice that for ordinary people overall.

We can contextualise these youth crime figures further by comparing them with the results for all UK *domestic* 'crime' stories in our sample. Of the 304 stories about domestic 'crime' in Britain, in 42 per cent (n = 127) the offenders/suspects were young people. For 'violent crime' stories specifically the figure was 50 per cent (96 of 192 stories about violent crime in the UK).[1]

Finally, young people seldom received notable attention in relation to politics and political activity. We coded for the subject 'Young people: Political Interest/Activity'. This code denoted either young people being shown to be active in connection with a political or social campaign, or discussion of the theme of young people and political interest or engagement. We found that it was a *'main'* subject in only 9 stories, or 3 per cent of stories with young people as a *'main'* focus (n = 286). It was slightly more frequent as a *'main/subsidiary'* subject, in 21 stories or 6 per cent of stories in which young people were a *'main/subsidiary'* focus (n = 358).

Young people as news sources

Young people received a total of 2646 seconds of speech as news sources. As Figure 5.1 shows, they were often depicted discussing the topic of

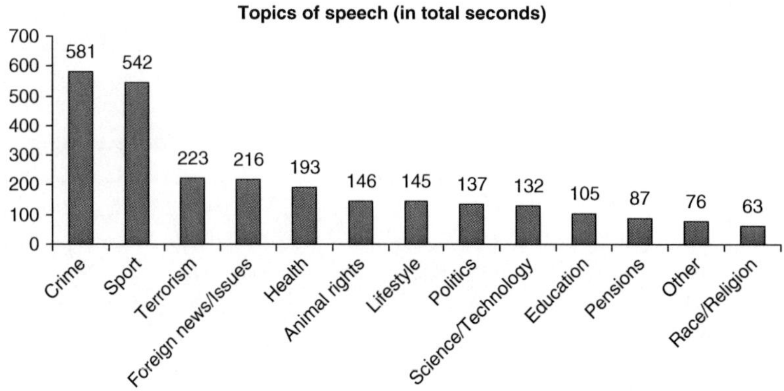

Figure 5.1 Young people as news sources

'Crime' (581 seconds). 'Sport' was not far behind (542 seconds), though over 90 per cent of this speaking time (498 seconds) came from clips of interviews with young football celebrities Rooney and Walcott (e.g. *Sky News at Ten*, Sky News 22.00, 9 May 2006; *ITV News*, ITV1 22.30, 2 May 2006). The third most frequently discussed topic was 'terrorism' (223 seconds). It is notable that Asian males were the only youth demographic that spoke on 'terrorism'. Taken together, 'crime' and 'terrorism' collectively accounted for 30 per cent of all seconds of speech by young people (804 seconds), sport comprised 20 per cent, politics 5 per cent (n = 137 seconds) and all social-economic issues combined 25 per cent (n = 668 seconds).

Gender and ethnicity of young people on the news

As 'story actors', we identified young people as being *'main'* in the 286 stories already discussed and as being *'main'* or *'subsidiary'* in 358 stories altogether. That is 17 per cent of the 2130 stories sampled. Young males appeared much more frequently than young females. They were in fact *'main'* story actors five times as frequently (206 stories compared with 41) and *'main/subsidiary'* more than three times as often (280 to 79). They also had much more speaking time than young females, comprising 70 per cent (1841 seconds) of all young people's speech (compared with 30 per cent for females, n = 805 seconds). Given that young people appeared mostly in stories about sport (especially football) and crime, and that in both kinds of stories the focus was almost exclusively on young males, this result is not altogether surprising.

'White' youths appeared on the news more frequently as '*main*' actors (n = 109 stories) than either 'black' (n = 77) or 'Asian' (n = 34) youths. This result held true for '*main/subsidiary*' appearances as well ('white', n = 162, compared with 'black', n = 117, and 'Asian', n = 63). 'White' youths also accounted for more than half of all young people's speaking time (52 per cent, 1377 seconds), with 'black' youths receiving 25 per cent (667 seconds) and 'Asian' youths 21 per cent (556 seconds). ('Unclear/Other' received 2 per cent, n = 46 seconds). More specifically, 'white' males were '*main*' story actors most frequently (n = 89 stories) of all youth demographics, followed by 'black' males (n = 68), 'Asian' males (n = 34) and 'white' females (n = 22). 'Black' females (n = 13) and 'Asian' females (n = 1) received the least attention. These trends continued for '*main/subsidiary*' as well. 'White' males also received the most speaking time (40 per cent, n = 1049 seconds), followed by 'black' males (16 per cent, n = 421 seconds), 'Asian' males (14 per cent, n = 366 seconds) and 'white' females (12 per cent, n = 328 seconds). 'Black' females (9 per cent, n = 246 seconds) and 'Asian' females (7 per cent, n = 190 seconds) were least prominent by this measure as well.

Again these results can be largely attributed to the focus on sport and crime. Half of the stories focusing on young 'white' males were about Wayne Rooney's foot injury (n = 45 stories) and, as we discuss momentarily, many of the stories focusing on 'black' males were about youth crime (which had a disproportionate focus on 'black' males, mainly as victims of crime).

Ethnicity of victims and perpetrators of violent crime

According to American research, crimes committed by black and Hispanic offenders tend to be reported most commonly while whites receive the most attention as victims of crime (Gilliam and Iyengar 2000; Klite, Bardwell and Salzman 1997). With this in mind, we coded for ethnicity of both young victims and young perpetrators in stories about young people and crime, focusing specifically on violent crime.[2]

Looking first at the 93 stories that focused on violent crimes committed *against* young people, we identified 67 in which the ethnicity of the *victim* was reported. (In 26 stories ethnicity was unspecified or the 'victim' in question was not a specific young person but 'young people' in general.) Of these 67 stories, 'black' youths were victims in 41, 'white' youths in 21 and 'Asian' youths in 5. (We should note that this does not include 9 stories involving an 'Asian' victim – the non-fatal stabbing of a 14-year-old schoolboy on 26 May – because none of these stories reported the victim's ethnicity, or his name.)[3] Turning to the 97 stories about violent crimes committed *by* young people, we identified 81 which related to a particular crime incident and 16 which dealt with the issue of 'young people and violent crime' more generally (e.g. a government knife amnesty targeted particularly at young people, and backgrounders on the issue of 'young people and knives'.) In the 16 stories about 'young people and violent crime' there was generally no particular emphasis on any one youth demographic (though implicitly the focus was invariably on young males). Of the 81 stories about specific crimes, the ethnicity of the young offender(s) was reported only in 29. This means that in two-thirds (64 per cent, n = 52) of stories about specific violent crimes committed by young people, the perpetrators' ethnicity was not reported. In the 29 stories in which ethnicity was identified, 'black' youths (n = 15) were perpetrators in more stories than either 'white' (n = 10) or 'Asian' (n = 8) youths. (Four stories featured both 'black' and 'white' offenders.)

Young people and knife crime

There were a number of reported incidents involving 'knife crime' during May 2006. Several of them received substantial media attention. Knife crime in fact comprised 60 per cent (n = 116) of all domestic (British) violent crime stories in our sample (n = 192). Two-thirds of these knife crime stories (65 per cent, n = 75) concerned young people as actual or potential offenders. Stories linking young people to knife crime made up well over half (57 per cent, n = 76[4]) of all stories about youth crime, domestic or foreign, in our sample (n = 134). They accounted for 78 per cent of reports about young people and violent crime (n = 97). As Figure 5.2 demonstrates, knife crime comprised a particularly large proportion of stories about youth crime from mid-May onwards.

Discussion

Our finding that television news reporting of young people tends to be most concerned with youth crime, particularly violent crime, and the vulnerability of young people (especially when the two are connected) lends support to Griffin's contention that youths tend to be represented 'as trouble' or being 'in trouble'. The only real exception to this was the 28 per cent of young people stories that focused on young celebrities, particularly football stars in the lead-up to the 2006 FIFA World Cup.

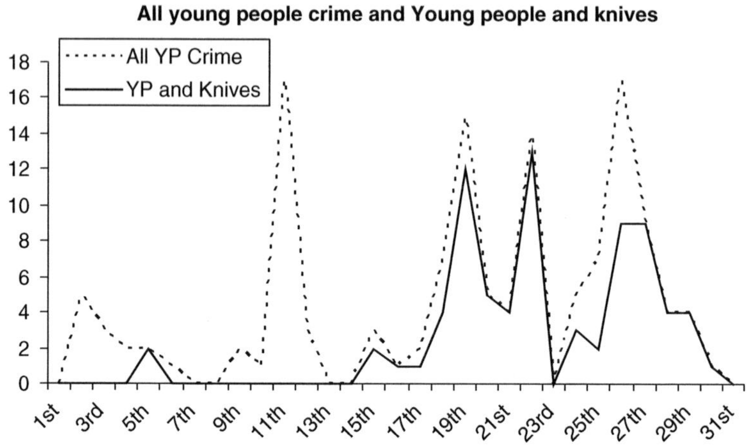

Figure 5.2 Frequency of 'young people and knife crime' stories, May 2006

Because of the focus on sport we did find fairly high levels of 'positive' coverage towards young people, but outside of sporting achievements there was very little attention given to successes or endeavours undertaken by the young. Taken together, the focus on crime (59 per cent) or celebrities (28 per cent) accounted for 87 per cent (n = 250) of all stories about young people in our sample. Our findings thus appear to reflect that television news tends to internalise the broader cultural dualism in the representation of young people, as either at the cutting edge of mass media spectacles, fashions and consumption trends, or law and order problems.

Our analysis of gender and ethnicity in violent crime stories provided some mixed results. Overall we did not find much evidence to suggest that British television news reproduces the ethnic biases towards white victims and non-white offenders that have been found on American (local) television news. Indeed, there were many more stories about attacks on black victims than either white or Asian victims. And, as is to be expected, the news treatment of victims was overwhelmingly sympathetic. This was particularly the case in the fatal stabbing of a 15-year-old schoolboy, Kiyan Prince, outside his school in North London (18 May 2006). This was the most reported youth crime incident in our sample (29 stories) and much of the initial (and most extensive) coverage contained emotional tributes to his wonderful personality and great promise as a student and athlete.

In terms of the young perpetrators of violent crime, black youths were again the most prominent demographic, though the statistical differences were not as great. The dominant trend was rather the non-reporting of offenders' or suspects' ethnicity (either in the dialogue or in the visuals). As a consequence, this part of the sample is too small to extrapolate any systemic trends in this regard. Similarly, though there was certainly a greater overall connection between black youths (males) and violent crime in our sample than for either white or Asian youths, whether as victims or perpetrators, it is difficult to form strong conclusions on the matter given that much of the connection arises from sympathetic coverage of black victims. However, on the question of gender, the dominance of crime and sport as topics in which young people featured had the effect of marginalising women as both story subjects and in terms of speaking time.

In our analysis we identified a high preponderance of stories about knife crime, most of which concerned young people. The problem of 'knife crime' became a steady news focus in the latter half of our sample following the stabbing of a female special constable on 11 May 2006. On some channels the coverage almost bordered on panic. After the Kiyan Prince murder, *Five News* was suggesting that 'knife crime' was 'sweeping the country' (C5 17.30, 19 May 2006), and following a reported 51 attacks over a 3-day Bank Holiday weekend (27–29 May)[5], *ITV News* said 'knife crime' was 'spiralling out of control' (ITV1 22.30, 30 May 2006). *Sky News at Ten*, *Channel 4 News* and *BBC News* all adopted more sober approaches, limiting their coverage primarily to straight reporting of the key incidents (though Sky did run three backgrounder pieces investigating young people and knives and the BBC had one analysis piece examining the use of metal detectors in New York high schools (19 May 2006)). An example of Channel 4's less sensational approach was their response to the Prince tragedy: 'As tributes are paid to the gifted teenage footballer Kiyan Prince, stabbed to death outside his school, police step up security in the area. But is there really a growing problem with people using knives?' (C4 19.00, 19 May 2006). Interestingly, however, perhaps the most subdued of all programmes was the children's news programme *Newsround*. While some of the other channels were warning of a 'knife culture that's out of control' (*ITV News*, ITV1 22.30, 30 May 2006), *Newsround* assured its viewers that despite a number of recent stabbings, two of which occurred just outside of schools, knife attacks were still rare. 'And remember that although these stories are very disturbing, knife attacks are still unusual, so try not to worry too much' (*Newsround*, BBC1, 26 May 2006).

Crime news and ideology

The recurrent alignment between young people and crime within television news which our quantitative analysis has found can be contextualised within a broader debate around the media representation of crime and its implications for public knowledge, debate and policy. Robert Reiner provides a succinct summary of the critique of the crime news agenda:

> media representations tend to exaggerate the threat of crime and to promote policing and punishment as the antidote. This is likely to accentuate fear, and thus support for law and order policies ... the media present viewpoints on crime and criminal justice policy which though not monolithic are loaded towards official perspectives. (Reiner 2002: 407)

The evidence for media exaggeration of crime is considerable. For example Schlesinger and Tumber found that 'violence against the person constituted 3.62 per cent of all notifiable offences reported by the police. However, such criminal acts comprised 24.7 per cent of crime-related items reported in the quality press, 38.8 per cent in the mid-market press, and 45.9 per cent in the popular press' (Schlesinger and Tumber 1994: 185; see also Coleman and Moynihan, 2003: 41; Graber 1980: 39; Maguire 2002: 339). Such frequently noted disjunctions have led some to claim that news coverage of crime is a major factor in shaping public (mis)perceptions of both the amount of crime which takes place in society, and the nature of that crime. The Home Office Working Group on the Fear of Crime, whilst by no means putting all the blame on the media for this state of affairs, nonetheless argued that it played a central role in creating it: 'Television, radio and the press feed the public much of their information about – and images of – crime. The pressure to increase circulation or attract audiences leads to simplification, over-dramatisation and sensationalism' (Home Office Standing Conference on Crime Prevention 1989: 23–4).

Similarly, George Gerbner and his colleagues in the US have argued that television helps to construct a 'mean world syndrome' amongst audiences. Gerbner argues that crime in prime time programming of all kinds (fiction as well as factual) is at least ten times as rampant as in the real world. Thus 'the world of television cultivates exaggerated perceptions of the number of people involved in violence in any given week, as well as numerous other inaccurate beliefs about crime and law enforcement' (Gerbner et al. 1986: 28). The cultivation of a 'mean world syndrome' has the potential of increasing a population's dependence on authority and their support for increasingly authoritarian policies. Media consumers 'may accept and even welcome repressive measures such as more jails, capital punishment, harsher sentences – measures that have never reduced crime but never fail to get votes – if that promises to relieve their anxieties. That is the deeper dilemma of violence-laden television' (Gerbner 2003: 345).

There may be some questionable assumptions around the uniformity of media effects in Gerbner's work, but there is evidence from the British context that suggests a link between media coverage of crime and public fears about crime. A recent *British Crime Survey* by Walker et al. found that the total number of crimes estimated by the survey has fallen in recent years, and that since 1995 violent crime in particular has fallen by 43 per cent. Yet a remarkable 63 per cent of people thought that crime in the country as a whole had increased in the previous two years, with 30 per cent believing that it had risen 'a lot' (Walker, Kershaw and Nicholas 2006: 34). Interestingly, usage of different media appears to correlate significantly with different perceptions of crime. Thus the survey found that readers of national tabloid papers were more than twice as likely to think that the national crime rate had greatly increased than those who read broadsheet papers (ibid.: 35). Reiner (2002: 384–5) argues that television news more closely resembles a tabloid news agenda than a broadsheet one, and certainly our statistical findings presented above would lend credibility to that view.

Five News special reports: Competing frames

Following the fatal stabbing of special constable Nisha Patel-Nasri outside her home on 11 May 2006, *Five News* did a four-part series of special reports into knife crime on 15, 16, 17 and 18 May 2006. We have chosen these reports as our case study because they focus on knife crime and they reproduce the close association between (knife/violent) crime and young people representative of a broader trend from the sample analysed. As noted earlier, almost two-thirds (65 per cent) of the 'knife crime' stories in the sample were focused on the issue of young people and knives. Yet, while typical in terms of its concern about young people and violent crime, the case study is also indicative of the modest range of differentiation within news provision in terms of formats. Among the 65 separate reports dealing with young people and knives there were 14 stories that did not deal with a specific incident but looked more widely at what *Five News* called 'the growing problem of knife crime in Britain' (C5 1730, 15 May 2006). Half of these were on *Five News* (n = 7), with ITV (n = 3), Sky (n = 3) and BBC (n = 1) also analysing the topic via 'backgrounder' news pieces.

The *Five News* special reports lasted between 3.5 and 5 minutes in length. As 'backgrounders' they were to some extent able to uncouple themselves from the event-driven nature of most news coverage (Ericson, Baranek and Chan 1991: 8; Reiner 2002: 386). The greater allocation of time allowed the reports to discuss potential causal factors behind crime while the step back from the daily news agenda also allowed a greater variety of voices and perspectives to be recruited into the reports. This includes the innovation of recruiting a lay reporter for one of the special reports. A key issue with regard to the case study is what happens to the loading of crime reporting towards official perspectives that Reiner complains of, when a greater range of voices and perspectives is recruited for formats outside the temporally more limited and arguably more standardised coverage of the news bulletins?

Interestingly, across the four *Five News* reports only one police officer (a DCI in the British Transport Police) and only one politician (a government minister) were interviewed, in reports two and four respectively. This represents a broadening beyond the 'official accounts' that usually dominate crime news (Chibnall 1977: 39). The other interviewees are spread across various types. The increasing role of victims within the criminal justice system has also been reflected in the media (Reiner 2002: 392). Thus the special reports give a prominent place to parents or relatives of knife attack victims, who feature in three out of the four reports. Former criminals, young people, youth club workers and a Black British rap group promoting an anti-weapons message also feature in the reports.

Giving some editorial control to either well-known figures and celebrities or occasionally ordinary people has become an interesting component of broadcasters' strategies to reconnect with audiences disaffected with the news agenda (Kevill 2002). Thus the final report in the *Five News* series is handed over to a 'lay' reporter with particular experience of knife crime, namely Frances Lawrence, whose husband Philip, a head teacher, was fatally stabbed by a young person outside the gates of his school in 1995. Her report crystallises particularly sharply the internal dissonance between two competing frames within the news discourse over the four reports. Media frames are interpretive grids that 'define problems ... diagnose causes, make moral judgements and suggest remedies' (Entman 1993: 52). First, there is a dominant framing of the issues within a law and order agenda that implicitly calls for top-down solutions to a problem that is understood in terms of criminal justice law and punishment. This frame was evident in many news reports around the issue of knife crime. For example, an *ITV News* report ends with the mother of a teenage knife victim issuing a plea for tougher sentencing for people caught carrying knives (*ITV News*, 26 May 2006).

Within the *Five News* special reports there was also a subordinate but still present alternative frame that identifies questions of resources for young people and grass-roots initiatives as a possible means of addressing the problem of crime. This grass-roots frame is unusual in crime news because it is problematic for the ideology of the dominant frame. It shifts attention to questions around the distribution of resources within society and implies that inequalities breed crime. Instead of already established criminal justice institutions accruing greater powers, solutions are sought elsewhere, not least within the groups from whom the ranks of criminals – of at least this type of street crime – are drawn. The question of resources and the development of strategies that include ordinary people in solutions to crime shifts the burden of responsibility for crime back onto society itself and its current structures, rather than locating it overwhelmingly as the responsibility of the criminal, as with the dominant law and order frame.

Statistics: Blinded by science?

As we have seen, statistics do not necessarily support the media view of violent crime as forever rising; yet the selection and presentation of statistics by television news is a recurrent feature of journalism. The second *Five News* report is a good example of the way news can use statistics to both lend credibility to its own reporting and inscribe a dominant law and order frame into it while still only interviewing one senior police officer within the report. It begins, as does the first report, with a series of statistics that seek to provide a rationale for the focus of the report(s): namely that knife crime is an urgent and escalating problem. Thus we are told that (1) 40,000 police officers have been threatened with a knife in the past two years. The news anchor who introduces the report presents this statistic verbally. The Police Federation is cited as the source for the statistic. (2) Youth knife crime is up 20 per cent on last year, according to a London-based report. (3) According to a MORI survey, 30 per cent of secondary school children admitted to having carried a knife. The last two statistics are contained within the report itself and are presented graphically against a still and moving image of a ferocious looking knife. It is the final statistic that provides the rationale for the first part of the report which conducts an experiment to 'test' the 'perception' (the reporter's words) that schools are 'awash' with knives. Thus, working with the London Metropolitan police (visually present but no representatives are interviewed), the report organises what it claims to be 'the first ever mass test for knives in a British school'. The viewer is shown the secret assemblage of a metal detector system in the blacked out school hall, followed by shots of the schoolchildren being screened later on. The result of this little narrative drama is actually quite revealing – at least potentially. Out of 200 children scanned, not a *single one* was found to have a knife or weapon of any sort. Yet, although the results of this experiment flatly contradicts the MORI survey quoted earlier, and although the reporter had spoken of putting the 'perception' of rampant knife carrying within schools to the 'test', there is in fact no return to or questioning of either public perception or the earlier quoted statistic. What follows are four brief interviews on the merits of conducting the experiment itself, with the clear implication that this might be something that could be institutionalised within schools. Of the three schoolchildren presented, two support the idea of randomly testing children for weapons in this way while a third questions it, suggesting that it is criminalising the students. The fourth interview (and thus arguably cancelling out the young person who raised questions about the experiment) is with the head teacher himself who while not advocating the institutionalisation of random testing *per se*, does suggest that this particular exercise helped send a 'strong message' to the children.

By focusing on the experiment (rather than the results, which cut against the public perception of a problem that is in some way out of control) and the possibility of extending it on a more permanent basis within British schools, the *Five News* report reproduces the dominant frame of 'law and order'. Here institutionalised surveillance is proffered as a potential answer to a problem, the pervasiveness of which is, on the evidence the report itself has presented, open to debate.

Although the police are not interviewed in this part of the report, the image of testing the children is visually reproduced later on in the report where we join a stop and search checkpoint using a metal detector in a Birmingham train station. Here a Detective Chief Inspector (DCI) from the British Transport police is interviewed. Having failed to find an example of a link between young people and knives with the school test, the report here notes that 'the first arrest was a fourteen year old with a screwdriver' (note, not a knife). The DCI comments that this

Illustration 5.1 Constructing the law and order agenda?
Source: Courtesy of *Five News*.

does not surprise him. The viewer is then informed that one-quarter of people arrested with a weapon are youths. Yet, if this statistic is designed to secure the conflation the series of reports seem to want to make between youths and knife carrying, then we are still left with *three-quarters* of arrests for carrying a weapon occurring outside the category of 'youth'.

Contradictions between visual and audio messages

The television news report does not completely ignore the results of the school experiment, however. It uses the results (no schoolchildren carrying knives) to link in another written report (cited but not named) that found that it was children excluded from school who were more likely to end up carrying knives. These are 'outcasts like Jason who ended up in gangs'. Here we have an interesting contradiction between the verbal message of the reporter in this transition to another scene and the visual presentation of and subsequent interview with Jason. While the reporter seems to be indicating some sort of system failure whereby excluded children are pulled into the ambit of a gang culture, the visual language and subsequent reporter language plays up the threat, menace and danger which Jason used to pose to society. He is introduced in long shot talking to the reporter under a bridge, with both figures in silhouette. The reporter tells us that 'a few years ago you wouldn't want to meet this guy down a dark alley'. The subsequent interview focuses sensationally on the use of knives and (with Jason leaning casually against the wall) the need to dispose of knives if they have blood on them. Thus the focus shifts away from a system failure (briefly alluded to) and onto the threat posed by an *individual* (Jewkes 2004: 45–6). The implication once again is that criminal justice solutions rather than reform of educational provision for young people are the answer.

We can contrast this presentation of Jason – who is young and black – with the presentation of another former criminal in the first report. Here Bobby Cummings – who appears to be in his sixties and is white – is introduced as a former hit man and bank robber. He, however, is shown as someone who is constructively redemptive, as he visits places like youth clubs to talk to children and young people about the dangers of carrying weapons and getting involved in violence and crime. It is hard to avoid the conclusion that the racial *and* age profile of Jason and

Illustration 5.2 The rhetoric of menace
Source: Courtesy of *Five News*.

Bobby are significant factors in the very different presentation they are given across the two reports.

The contradiction between the verbal and visual components of the media message which we find in the presentation of Jason are also evident in the third report of the series which dedicates itself to finding out why 'so many' (according to the news anchor) young people are arming themselves. The report cites a study conducted by academics at Glasgow University that explored the motivations for knife carrying. Here the report cuts to footage of young people walking about busy streets. The reporter summarises some of the findings of the academic study, which suggests that young people may carry knives for (a) status acquisition (b) fashion reasons (c) to intimidate others (d) as protection. But as each of these reasons are cited, the news report cuts to footage of young people walking about busy streets (framed anonymously, below the face) and simulates an X-Ray effect, freezing the image of a young person, turning it into a negative and then digitally 'planting' as it were, a concealed knife on their persons, under their clothes or in a bag. Thus an argument about why young people may carry knives is dramatically and sensationally yoked to visual imagery suggesting that knife carrying among young people is widespread and routine. Again the effect is to imply that a criminal justice solution is urgently necessary, even though at least two and possibly three of the four reasons stated suggests that there are cultural factors at play that need to be addressed. The inclusion in this report of two (male) representatives from a British rap group with an anti-violence message, Big Brovaz, also suggests cultural factors as well (and balances an earlier part of the report which explored whether violent computer games might be to blame). In the interviews, both members of the group also identify a lack of material resources, facilities and opportunities for young people as causal factors in youth crime.

News frames and the lay reporter

The final report led by Frances Lawrence combines in one figure the trends towards including the voices of victims and their relatives in crime reporting and the trend towards handing over some sort of editorial control to guest presenters. The report also enlarges the grassroots framing of the issue of crime already invoked in the third report. Francis Lawrence begins by visiting the Stonebridge Estate and talks to young people about how they improved the area by putting up a football cage. A local youth worker is interviewed on site and insists that it is the young people themselves who are the solution to problems of crime. We are then introduced to Frank Knight who is another reformed criminal. Like Jason from the second report, Frank is young and black. But the representation of this former drug dealer is very different. Whereas the first shot of Jason in long shot talking to the reporter evoked visual cues that seem to inscribe him into a rhetoric of continuing menace, Frank is presented in long shot talking to Frances as they walk together down the road. Frank is stripped of any menace and the scene suggests reflection and 'moving on' from his past life. This lack of menace around Frank is reinforced further by the location of the next scene in which we actually hear their exchange: a children's playground. This is a very different visual contextualisation of Frank as compared with Jason. It may remind us that he was once a child, or imply a recovery of innocence, while the substantive content of the interview is also different. Whereas Jason was asked to discuss the pros and cons of using a knife, Frances asks Frank what he thinks of his life now that he has escaped involvement with drugs related crime. In short the lay reporter has introduced some significant modifications into the usual practices of television journalism.

The report then shifts geographically and socially away from the estate as Frances Lawrence visits the Home Office and interviews the Police and Security Minister, Liam Byrne, MP. Here the discourse also shifts away from the question of resources and local initiatives and towards a law and order framing of the issue. For example, the discussion focuses on 'stiffer penalties' for being caught in possession of a knife, making it more equivalent to carrying a gun. Reference is also made to the Violent Crime Bill making its way through the Houses of Parliament. Frances Lawrence's concluding summation (now back in her house) is riddled with the contradictory pull between these two frames. On the one hand she calls on the government to listen to and learn from the sort of young people who have appeared in the report and she mentions Frank specifically (which it is highly unlikely a professional broadcast journalist would do). On the other hand she calls for 'firm legislation' to confront knife crime. While theoretically it might be possible for the law and order frame and the grass-roots frame to be at least partly reconcilable, in practice, the former excludes the latter and finds in the grass-roots frame an uncomfortable reminder of the sort of questions that a narrowly conceived law and order frame cannot really address.

Across the four reports as a whole then, we find significantly contradictory meanings and possibilities. The reports seemingly unintentionally open up questions about their own use of statistics, while the conflation between young people and knife crime is both secured but also at times fraying apart. Visual and audio information sometimes contradict each other while the dominant hegemonic frame of law and order has to make some accommodation to a subordinate frame orientated towards grass-roots solutions and the issue of material resources. The reports, while perhaps more internally dissonant and contradictory than usual because they were 'backgrounders', were still operating within the ideological boundaries which shape the crime news agenda as a whole.

Conclusion

We can only properly understand the way television news represents young people by reminding ourselves of the broader social and political context within which these representations take place. We have situated television news and its role in cultivating a public sphere of debate and comprehension within the context of the rise of neoliberalism and the consequent dismantling of the welfare state and the social democratic compromise between capital and labour of the post- Second World War period. The negative impact of neoliberal policies on the fabric of society requires in turn a more punitive criminal justice system to deal with the fall-out from the social and economic dislocations caused by unleashing market mechanisms and imperatives at every level of society. Under the New Labour governments, the prison population has risen from 61,000 in 1997 to 83,000 in 2009, and government policy is predicated on increasing that figure still further. New Labour has created over 700 new offences since 1997, many of them focusing specifically on young people. As one report notes

> people are being criminalised for things now that their parents and grandparents got away with. Children are now dealt with more harshly than adults. There appear to be numerous instances where incidents that used to be regarded as normal adolescent behaviour fifteen or twenty years ago, are now being seen as low-level criminal activity. (Wood 2008: 10)

This amounts to the effective criminalisation of the working class, and the representations of young people on television news cannot be divorced from this context. We have to remind ourselves that the category of young people is deceptively 'universal', concealing or at least diminishing from consciousness the class dimension that is at work in these representations. But if we look at the majority of young people who appear on television news, look at their faces, their clothes, their gestures and language, then it is clear how significantly class intersects with 'youth' in the world of television news.

It is not good enough for broadcast journalists to say that they do not do 'good news' stories in general or that there are limits to what they can do within the temporal constraints of news bulletins. The news *could* flag up a broader range subjects related to young people's lives, even within the constraints of current formats. And journalists ought to have enough reflective powers to recognise the ideological implications of screening out wider social and economic factors in their presentation of crime news. Our research demonstrates that young people appear most commonly as either victims or perpetrators of crime and there is a strong association between young people and violent crime (specifically knife crime in our sample). As sources of comment, young people have a marginal place in news representations, and when they are used it is typically in relation to questions of crime. Constructed overwhelmingly as problems for 'law and order' the conditions would appear to be in place for young working-class people to become (and arguably to some extent they already are) the locus of moral panics. This 'symbolic criminalisation' fits young people into patterns of crime reporting that have long been critiqued as vehicles for exaggerating public fears and promoting increasing state power as a viable solution to crime. Our research also indicates that television news reproduces a broader cultural dualism in thinking about young people: 'positive' stories about young people are overwhelmingly linked to celebrities and therefore to media spectacles and consumer trends. The current agenda for television news is unlikely to make a positive contribution to the formation of public opinion or public policy as far as young people are concerned, let alone make television news more attractive to young people.

6
The Monopolisation of Political Discourse

This chapter explores how television news and discussion programmes represent the world of 'formal' politics, which is to say the world populated by a professional class of political 'doers' organised within political parties, who make and implement policies within institutions that are supposed to be representative of and responsive to the public. Within a representative democracy, every voter is theoretically equal. But in practice, political power is separated from economic power. Voters vote for the former. The latter is unelectable and largely unaccountable to citizens. The story of the rise of neoliberalism is the story of how political power has become ever more subordinated and integrated into the preferences and imperatives of economic power as the social democratic/welfare state structures developed in the middle of the past century are dismantled. Inevitably, this process affects the structure of politics and the representation of that structure to the public via the media. For example, the class of political 'doers' operate increasingly within new party-type organisations. In the social democratic period, parties had a traditional mass base of members and were orientated towards constructing a hegemonic constituency of voters; today these ostensibly ideology-free structures operate on a declining membership base and are internally structured to minimise debate and dissent and orientate themselves towards voters as consumers, selling technocratic solutions rather than espousing ideological differences with opponents (Mair 1997: 34–8). For voters the choice is often one between technocratic managers and their 'brand' identities. A similar double process of building hierarchical structures *within* political organisations and consumerisation vis-à-vis the voter/public/citizen can be found within television coverage of politics. In this chapter we provide a statistical and qualitative analysis of the hierarchical coverage of politics by UK television news. We find that there is a rigidly structured hierarchy of political access and focus, whereby the Prime Minister dominates over the cabinet, the cabinet dominates over ordinary MPs, the governing party dominates over the opposition, the three main parties dominate overwhelmingly over smaller parties, and the political elites dominate over ordinary members of the public. But this hierarchical focalisation (i.e. our way into the political narrative), which testifies to the media's deference to *political power,* is combined with coverage that testifies to the media's subordination to the imperatives of the market, that is *economic power.* We shall show that the framing of television news is skewed towards 'horse race' and personalisation coverage both of which outweigh 'policy' issues. Such skews may produce bitter conflicts between the political elites and the media. But this does not signify in any way that the media are interrogating the neoliberal *policy consensus* amongst elites. Media interrogation of political elites remains resolutely *anti-systemic, personalised or focused on processes or means, not an interrogation of ends.* In the context of the political consensus amongst elites, declining legitimacy for formal political institutions and an ageing profile for television news, the interlocking relations between the political establishment and television news on the one hand, and the commercial imperatives driving news in the direction of personalisation over policies on the other, is antithetical to the democratic process.

Monopolisation and spectacle

One institutionalised expression of the interlocking relations between the political class, the broader state apparatus and UK television providers in the UK can be found in Millbank. These are the television studios opposite the Houses of Parliament that give the broadcasters a direct conduit into Westminster life and actors. As Steve Anderson, executive producer at Mentorn for BBC1's *Question Time* notes, this is a 'machine' that needs feeding:

> I think one of the problems here is you create a machine to service the beast and then once you've created this machine you have to keep using the machine otherwise why has the BBC, ITN and Sky got huge resources down at Millbank if they're not going to use them? So they become a self-fulfilling prophecy in the end. The resources are there so they have to produce. But if we're spending money there we're not spending money in Leeds or in Darfur or wherever. So when your decision is taken that's where you're investing, that's got to produce news.

This physical proximity towards formal political authority also bleeds into a convergence of values and ideas. The US media critic Lance Bennett has described this in his 'indexing thesis':

> Mass media news professionals from the boardroom to the beat, tend to 'index' the range of voices and viewpoints in both news and editorials according to the range of views expressed in mainstream government debate about a given topic. (Bennett 1990: 106)

The priority accorded to elected representatives in the media assumes that the electoral institutions are doing a reasonably effective job of representation. Yet within a UK context, this has always been problematic. As Marsh, Richards and Smith (2003) argue, the culture that permeates the structures of British government is that political representatives are accountable to the public at periodic elections but not particularly responsive to them in between elections, thereby sidelining popular participation and inclusion in political processes (312). In the context of globalisation, disjunctures between political representatives (especially those political representatives with policy-making powers) and public opinion are likely to become wider and more routine. Television news actually finds it very difficult to explore the relationship between political representatives *and* the public those representatives are supposed to represent. Here is Nick Robinson, the BBC's chief political correspondent rationalising his particular job:

> A political editor is reporting on power, it's reporting on political power and political power resides in Westminster... it doesn't reside anywhere else. So predominantly the coverage will be driven by Westminster. Do I sound like a spokesman? Got to be careful not to sound like that. Now, politics moves as a result of the movement and opinion of people outside Westminster and it's certainly arguable that we could do better at trying to reflect that, whether it's the anti-war protests... there was some frustration [during the anti-war protests] from part of the audience saying, well, why weren't you with us on the streets saying look at this sort of public opinion? Now, we should have been on the streets, and we were and we did report the process but perhaps the overall coverage could have done more to reflect that. But in the end I report on power and those people didn't have it and he [Tony Blair] did and that was the correct reporting of who had power. And they wouldn't have had power just simply because I reported on them, by the way. You know, if only I'd

reported on the marches more that wouldn't suddenly have changed the position. They still wouldn't have had power.

It is interesting that Robinson sees his job in terms of reporting on formally constituted *power*, rather than representation, an indication perhaps of the declining plausibility that our political representatives are all that representative. This pragmatic framing of political action in terms of power, however, raises problems both for Robinson's position and the relationship between that political power and its supposed foundation: the people. Robinson is aware that reporting on power might make him look like a spokesperson *for* power. This might be particularly the case the more television news focuses on the Prime Minister and the upper echelons of political power, where the reporter becomes reliant on favourable access to sources and off the record briefings ('sources tell me ...'). Robinson is also aware, theoretically, that 'politics moves as a result of the movement and opinion of people outside Westminster'. At the same time he feels that greater coverage of this 'movement and opinion' in the case of the war on Iraq would not have changed the fact that political power rested with the Prime Minister.

Illustration 6.1 Nick Robinson, reporting on political power
Source: Courtesy of *BBC News*.

Yet, although formal political power does, of course, rest with the Prime Minister, the movement and opinion of people (against the war) is also a political power, less formal, not constituted in the rituals of the state, but political, representative and democratic nonetheless. And while in the short term Tony Blair was able to exercise political power as Prime Minister, his political authority and power started to decline and never really recovered from the moment he took the UK into the war on Iraq. It was opposition to that war, largely coordinated via the Stop the War Coalition, that drained the Prime Minister's political power in the medium and long term. And yet it is precisely this half of the equation (the 'movement and opinion' of people outside Westminster) that, as we shall see, is largely missing from the coverage of politics. The broadcasters do not have the frameworks, the initiative, the routine practices to deal with this 'movement and opinion' of people – in whatever forms of organisation – outside Westminster. The broadcast media's conception of 'civil society' – a concept denoting a plurality of organisations standing between the state and the private individual – is exceptionally thin and underdeveloped. This hierarchical structure means that the representativeness of the prioritised political actors is *assumed* by television news and not tested or explored. Within academic analysis of television news, the marginalisation of the public (Brookes, Lewis and Wahl-Jorgensen 2004) and the narrow focus on political elites (Philo 1995: 198) has long been associated with a deferential attitude on the part of the media towards political elites. However, recent trends within the UK media scene appear to be moving away from deference towards what has been called cynicism or populism. Such populism is only apparently in contradiction with the hierarchical structure to the coverage of political discourse which we find here. Crucially, this new modality of cynicism and populism in no way breaks with the *policy consensus* of neoliberalism. Indeed, it helps to cement it further in place by inculcating into the public the idea that politics is a remote spectacle of drama or corruption, thus naturalising the deregulated market as *the* place and mechanism for satisfying our needs. That contemporary 'critical' coverage of politics is focused mostly upon 'horse race' (popularity), strategy, scandals, gaffes and personalities, rather than the more substantive issue of *policy*, is a trend that has been proven time and again by media researchers. The American research is particularly thick with studies demonstrating the subjugation of policy issues to such foci (e.g. Brewer and Sigelman 2002; Cappella and Jamieson 1996; Deardourff 1996; Farnsworth and Lichter 2004; Gan, Teo and Detenber 2005; Kerbel, Apee and Ross 2000; Lichter 2001; Norris 2001; Patterson 1994).

As Eilders et al. (2004) observe, 'Events with certain characteristics like the status of the actors, personalisation, negativism and conflict are regarded as particularly newsworthy' (221). Recent research points towards the growth of these traits within British news as well. For instance, an analysis by Goddard, Scammell and Semetko (1998) into television coverage of the 1997 general election campaign found that, though 'due attention was paid to substantive issues', there was 'a considerable increase in stories addressing campaign conduct and strategy' (171–2). Similarly, subsequent studies of the 2001 and 2005 general elections by Loughborough University's Communication Research Centre have reported that 'coverage of the electoral process itself (i.e. the actions, strategies and prospects of the participants) was the most prominent topic in election coverage by a considerable margin' (Deacon et al. 2005: 25) and that 'the majority of items in all national media sectors contained either no or negligible descriptive policy-related information' (ibid.: 30). The preponderance of what some scholars call 'horse race', 'strategy' or 'game' framing (e.g. Cappella and Jamieson 1997; Patterson 1994), is feared to provide citizens 'with little or no information on which to base an intelligent voting decision' and 'tends to make voters spectators rather than participants in the electoral process' (Deardourff 1996: 106). There is also evidence to suggest that it is not the kind of focus most viewers actually want. Research commissioned by the Electoral Commission after the 2001 British general election campaign identified that 'most people felt that the coverage focused too much on the leaders and their personalities, and the campaign process' (2001: 55). This result 'echoed the wider findings of the opinion pollsters that the public wanted more information on policies and what the parties stood for, and less of the political gossip, campaign tactics and personality characterisation' (ibid.: 55–6). If this is what the public wants then political coverage by television news and political communication by the political parties themselves appear to be going in exactly the opposite direction. There has been considerable debate about the tendencies towards 'presidentialism' within British politics, especially since the successive elections to government of New Labour, since 1997. Although the term 'presidentialism' is ultimately inadequate as a critical category and inappropriate when applied to the British political system, it does capture something of the way political leaders are becoming differentiated from their party bases in terms of media attention and promotion (see Foley 2004: 293; Heffernan 2003: 354 and Bevir and Rhodes 2006: 681). But the personalisation of the political, of which 'presidentialism', horse race and scandals are all interlinked, has also rebounded on the political class in terms of reportage that has become increasingly willing to criticise politicians within the narrow terms outlined above. This has led a number of writers to argue that the media is to blame for public disenchantment with the political process which is essentially populated by honest men and women trying to find workable solutions to difficult problems (Barnett 2002; Cappella and Jamieson 1997; Lloyd 2004). This both uncritically accepts the media's view of itself as an independent fourth estate (before its decline) and cuts analysis of the political sphere off from any broader understanding of the social and economic power relations within which political elites work. Thus, Barnett (2002) is able to identify that the intensifying commercialism of the British media market is one important factor in the shift towards 'cynicism', along with an overreaction by journalists towards the professionalisation of political communication management by the parties themselves (401–3). But he is unable to identify that the very political elites he wants to throw a protective shield around are responsible for the policies which have made the media profit hungry and intensely competitive. Nor can he link those policies in turn to the broader globalisation of neoliberalism (see in contrast Leys 2001).

Methodological considerations

This chapter's analysis draws from the same sample of 2304 news stories discussed in Chapter 5.[1] Much of the analysis also focuses more narrowly on a smaller subsample of stories specifically about 'politics'. This subsample comprises all stories in our main sample in which 'British Politics' was coded as a *'principal/main'* story subject (n = 517 stories). The results we provide below are also more detailed than those presented in Chapter 5. For our analysis of 'story actors' and 'news sources', we subcategorised 'British politicians' by political party (e.g. Labour, Conservative, Liberal Democrat, Green, BNP), and within the three largest parties (Labour, Conservative, Liberal Democrat) into two groups to signify 'cabinet' ministers and 'non-cabinet' ministers. (We determined the 'cabinet' or 'non-cabinet' status of party officials using contemporaneous lists released by the parties themselves on their official websites (accessed May 2006).) For our analysis of 'story subjects', we subcategorised 'British politics' into variables including 'Local Government', 'Regional Assemblies' and a number of political 'news frames' (which we explain momentarily). It was possible to code multiple 'subjects', 'actors' and 'news sources' in a single story and each subject was given a prominence ranking (*'main'* or *'subsidiary'*) as outlined in previous chapters.

As part of our more detailed analysis of story subjects, we investigated the relative prominence of different political 'news frames'. Drawing upon American studies, we coded for the existence of three broad news frames: 'horse race/strategy', 'competence/integrity' and 'policy issues'. We defined the first frame, 'horse race/strategy', in line with Kerbel, Apee and Ross' (2000: 12) definition that such coverage asks viewers to experience politics 'as a contest with political winners and losers while detailing the actions they take in the pursuit of victory' (or power, status). Within this category we included stories that emphasised election contests, political conflict generally (either between or within parties), strategy, tactics, popularity (e.g. opinion polls), or the status of a party or individual politician (e.g. who's in or out, up or down). This frame also largely corresponds to what has been labelled 'game' framing in some studies (e.g. Patterson 1994).

The second frame, 'competence/integrity', covers stories in which emphasis is given to the abilities, performance or trustworthiness of politicians. This can be positive or negative emphasis. Such coverage has been both praised as an example of fourth estate or watchdog journalism, and criticised for unfairly souring public attitudes towards politicians and the democratic system. In the latter case it is argued that coverage too often degenerates into 'feeding frenzies' of questionable public value (Sabato 1991). We defined our final frame, 'policy issues', in line with several US studies (e.g. Brewer and Sigelman 2002; Gan, Teo and Detenber 2005). It covers stories that emphasise policy details, ideas or outcomes, or that give background or context to key policy debates. Coverage of this nature is widely seen as among the most substantive forms of political coverage because it looks beyond hoopla, point scoring and controversy to address politics' impact and consequences on individuals and society. Finally, we have complemented our statistical analysis of news programmes with statistics drawn from a less detailed analysis of political discussion programmes aired in our sample: 4 episodes of *Question Time* (BBC1, Thursdays 22.30), *Sunday AM* (BBC1, Sundays 09.00) and *Sunday Live* (Sky News, Sundays 10.00), 3 episodes of *This Week* (BBC1, Thursdays 23.35) and *The Politics Show* (BBC1, Sundays 12.00), 1 episode of *Jonathan Dimbleby* (ITV1, Sundays 10.30) and 19 episodes of *The Daily Politics* (BBC2, weekdays 12.00). This encompassed every episode of these programmes that appeared in May 2006. The analysis of these seven programmes incorporated a quantitative analysis of news sources. This was done in exactly the same way as for the news sample, enabling direct comparisons. Through a more qualitative critique we can also compare the subject agenda(s) of these seven political programmes with those of the analysed news programmes. Through this we can investigate the extent to which these additional hours of political programming enable attention to be given to issues outside of those already covered on the news, or whether these programmes offer rather more of the same. We will also briefly consider differences in how politics is 'framed' across the two formats.

Results: The representation of politics on British television

Our coding period (1–31 May 2006) included the last four days of an election campaign (1–4 May) for the local government elections that took place on 4 May 2006. With the possible exception of an additional few days that included a significant amount of post-election analysis (5–8 May), the rest of the sample was comprised of non-election political news, a topic rarely explored in statistical surveys of political coverage. We found that across our news sample, British 'politics' was identified as being a '*main*' subject in 517 stories, or 22 per cent of the full sample of 2304 stories (see Table 6.1). Of these stories, 97 per cent (n = 494) focused on Westminster events, issues and personalities, compared with 5 per cent on local government (n = 27) and 2 per cent on the regional assemblies (n = 10).

Of the programmes in our sample, *Newsnight* had easily the highest percentage of 'politics' stories. Indeed, it had more than twice the six-programme average (52 per cent). The level of political coverage was relatively consistent across the five other main news bulletins (between one-fifth and one-quarter of stories), except for *Five News*, which was well below the average (only 13.5 per cent). *Newsround* had very few stories about British politics.

Table 6.1 Percentage of stories with UK politics as a '*principal/main*' subject

Programme	Political stories	All stories	Per cent
BBC News	87	366	23.8
Newsnight	115	220	52.3
ITV News	67	324	20.7
C4 News	115	523	22.0
Five News	46	340	13.5
Sky News	78	381	20.5
Newsround	9	150	6.0
Total	517	2,304	22.4

If we look at the total proportion of all *news time* taken up by the stories with 'politics' coded as a 'main' subject, we find significantly higher percentages across all programmes. Together they took up 31 per cent of all news time in our sample (n = 84,196 out of 269,966 seconds of story time). This reflects that stories dealing with politics tended to be longer than average. Indeed, almost half of all lead or 'first headline' stories in our sample, which tend to be covered in the most depth, had politics as a main focus (91 of 197 bulletins, or 46 per cent).

Table 6.2 shows results for the number of stories in which politicians aligned to different British political parties featured as either *'main'* or *'main/subsidiary'* story actors.

Labour politicians and spokespersons featured much more frequently than representatives from other parties. As *'main'* story actors, they featured in seven times as many stories as Conservatives, who were more than twice as prominent as Liberal Democrats. The minor parties featured in fewer stories again. This imbalance was also clear in the attention given to party leaders, with Tony Blair featuring as a *'main'* focus (n = 161 stories) in 40 times as many stories as Menzies Campbell (n = 4 stories), and 9 times as many as David Cameron (n = 17 stories). Blair was in fact a 'main' focus in almost twice as many stories as all Conservative and Liberal Democrat story actors combined (n = 81 stories). Among the minor parties, British National Party (BNP) representatives featured most frequently, though *'main'* in only 8 stories. Respect Party MP, George Galloway, was a *'main'* actor in 4 stories (he was the only Respect representative in the sample), while the Green Party was 'subsidiary' in only 2 stories and 'main' in none. Political actors from parties other than those mentioned (including independents and

Table 6.2 Political actors as *'main'* and *'main/subsidiary'* story actors: By party

Party	Main	Main/Subsidiary
Labour	439	569
Conservative	60	170
Liberal Democrat	24	67
Respect	4	4
BNP	8	11
Green	0	2
Other Party/Independent	7	18
N. Irish/Scottish/Welsh Assembly	8	10
Local government	13	34

Table 6.3 Prominence of 'cabinet' and 'non-cabinet' story actors: By party

	Cabinet MPs		Non-cabinet	
Party	**Main**	**Main/Sub.**	**Main**	**Main/Sub.**
Labour	307	419	88	189
Conservative	30	102	17	45
Liberal Democrat	11	37	5	13
All three parties	327	445	106	226

cross-bench peers) were coded as *'main/subsidiary'* in 18 stories. Most of these stories related to either the recurring topic of pensions reform (Lord Adair Turner) or the voting down of a euthanasia bill on 12 May (multiple Lords).

Within the three 'major' political parties (Labour, Conservative, Liberal Democrat) we also coded for the prominence of 'cabinet' party officials compared with 'non-cabinet' representatives (e.g. non-cabinet MPs, MEPs, Lords, other party activists or spokespersons). As detailed in Table 6.3, 'cabinet' officials were 'main' or 'subsidiary' story actors more often than their 'non-cabinet' counterparts, across all three parties. Cabinet MPs from any of these parties were *'main'* (n = 327 stories) in three times as many stories as non-cabinet MPs (n = 106). They were twice as prominent at the level of *'main/subsidiary'* (n = 445 compared with 226 stories).

Hierarchy of news sources

Across the full sample of 2304 stories, 'British politicians' contributed 29 per cent of the total speaking time by all news sources (n = 23,322 out of 79,716 seconds of speech). Table 6.4 depicts the breakdown of political news sources by party.

Labour representatives received far more speaking time than any other party and accounted for 66 per cent of the total speaking time for all political actors. Conservatives had almost four times less speaking time than Labour and contributed 18 per cent of political actors' speech. Liberal Democrats accounted for 10 per cent. Among the party leaders, Tony Blair (n = 2096 seconds) had nearly three times as much speaking time as David Cameron (n = 768 seconds) and more than five times as much as Menzies Campbell (n = 404 seconds). The combined total for all other (minor) parties and independent MPs or Lords was only

Table 6.4 Speaking time for political news sources: By party

Party	Seconds	Per cent (n = 23,322)
Labour	15,489	66.4
Conservative	4,205	18.0
Liberal Democrat	2,275	9.8
Other Party/MP/Lord	708	3.0
Other Political	645	2.8
Total	23,322	100.0

3 per cent of speaking time for political actors. Within the three main parties (Labour, Conservative, Liberal Democrat), speaking time was fairly evenly spread between cabinet MPs and other, non-cabinet representatives. Cabinet MPs accounted for 54 per cent (n = 11,782 seconds) of speaking time for all Labour, Conservative and Liberal Democrat politicians (n = 21,969 seconds). (Non-cabinet sources received 10,187 seconds of speech.)

The public in politics stories

In addition to analysing news treatment of 'political' story actors (as above), we examined the representation of other social actors in news about (formal) British politics. We paid particular attention to the representation of 'ordinary people', or 'the public'. We found that 'ordinary people/the public' were coded as being *'main'* story actors in one-quarter (27 per cent, n = 140 stories) of our 517 'politics' stories and *'main/subsidiary'* in half (50 per cent, n = 260 stories). These results suggest a considerable level of access for ordinary people in political news coverage; however, a much better indication of how far the public were able to define the political agenda is in the speaking opportunities they received. The public contributed only 9 per cent (n = 2786 seconds) of the total speaking time for all news sources within our 517 politics stories (n = 32,026 seconds). In contrast, politicians accounted for 62 per cent (n = 19,919 seconds) and 'experts' of various descriptions (academics, former officials, researchers, scientists, journalists, writers, doctors and lawyers) received 16 per cent of speaking time (n = 4980 seconds). The comparatively low percentage for 'the public' reflects in large part the fact that while they were present in many political stories, often they were present only as an abstraction (e.g. 'public opinion') or a topic of interest (e.g. 'immigrants', 'voters', 'foreign prisoners'). It also reflects that the public had much shorter speaking opportunities than other actors. Comments by ordinary people were usually short sound bites (often vox pops), while politicians and experts tended to receive longer sound bites and to be interviewed in the studio or via satellite. Hence the average speaking time for ordinary people was only 12 seconds, compared with 34 seconds for politicians and 42 seconds for 'experts'.

'Young people' specifically were coded as *'main'* actors in 4 per cent (n = 19) of politics stories and *'main/subsidiary'* in 9 per cent (n = 44). They spoke in only 4 per cent of stories (n = 20), and altogether contributed only 1 per cent (n = 353 seconds) of speaking time for news sources.

News frames

Over the full sample, our analysis of political 'news frames' showed a higher frequency of stories emphasising 'horse race/strategy' and 'competence/integrity' than stories focusing on 'policy issues' (see Figure 6.1).

As *'main'* story subjects, 'horse race/strategy' (n = 209 stories) and 'competence/integrity' (n = 212 stories) featured respectively in 40 per cent and 41 per cent of the 517 political stories. Both were more common than 'policy issues' (26 per cent, n = 133). However, the distribution of the different frames was uneven across the sample, with most of

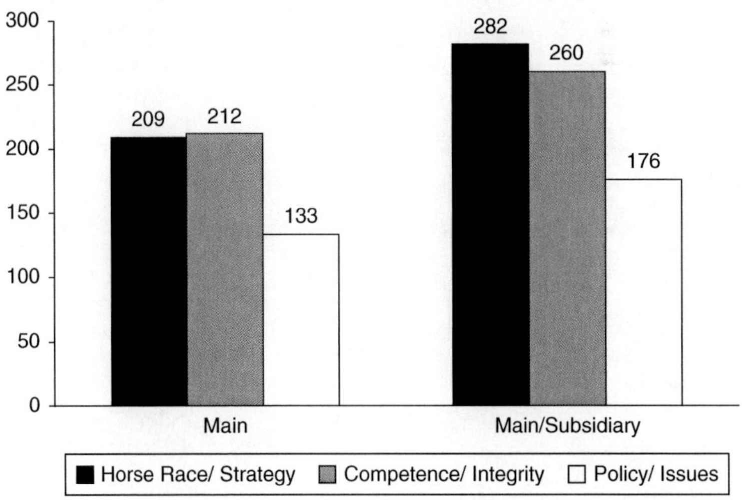

Figure 6.1 Stories with political news frames as *'main'* or *'main/subsidiary'*

the 'horse race/strategy' stories appearing in just the first eight days of the sample period (1–8 May), in connection with local government elections on 4 May and a cabinet reshuffle for Labour on 5 May. In this eight-day period, 'horse race/strategy' was '*main*' (n = 117) twice as often as 'competence/integrity' (n = 63) and ten times as often as 'policy issues' (n = 11 stories). Thereafter (9–31 May), it was less frequent (n = 92 stories) than either 'competence/integrity' (n = 149 stories) or 'policy issues' (n = 122 stories).

Political discussion programmes

Our analysis of news sources in political discussion programmes revealed some close similarities between these programmes and the political news coverage. The programmes we looked at were *Question Time, Sunday AM, Sunday Live, This Week, The Politics Show, Jonathan Dimbleby* and *The Daily Politics*. This encompassed every episode of these programmes that appeared in May 2006. Politicians were again the dominant group, comprising 56 per cent of speaking time for all sources quoted. 'Experts' were again second, accounting for 19 per cent, while the public contributed only 8 per cent (see Table 6.5). Excluding the weekday programme *The Daily Politics*, which aired much more frequently (19 episodes) and could therefore potentially skew the overall results, we find only slight variations to the key percentages. Politicians and experts still dominated discussion (58 per cent and 17 per cent of speaking time respectively) while ordinary people received comparatively few speaking opportunities (7 per cent of speaking time).

Between political sources specifically, there was a bit more balance on the discussion programmes than on the news, at least between Labour and Conservative. Across the seven programmes, Labour politicians

Table 6.5 Sources in political discussion programmes, May 2006

Sources	**Seconds**	**Per cent (n = 62,316)**
Politicians	34,850	55.9
'Experts'	12,115	19.4
Public/'Ordinary people'	5,162	8.3
'Celebrities'	4,240	6.8
International leaders	2,042	3.3
Others	3,907	6.3
Total	62,316	100.0

contributed 55 per cent (n = 19,107 seconds) of politicians' speech and the Conservatives 32 per cent (n = 11,081 seconds). The Liberal Democrats accounted for 10 per cent (n = 3410 seconds), all other, 'minor' parties 3 per cent (1079 seconds), and local government councillors 0.5 per cent (173 seconds). Among politicians from the three main parties, 'cabinet' officials actually received less speaking time than non-cabinet representatives across the seven programmes, by 48 per cent to 52 per cent (16,171 to 17,427 seconds). However, this result is significantly skewed by *The Daily Politics*, which screened much more frequently, and on which non-cabinet party figures made up 70 per cent of speaking time. Excluding *The Daily Politics* and looking just at the weekly programmes, cabinet officials (57 per cent, n = 12,856 seconds) received more speaking time than non-cabinet (43 per cent, n = 9729 seconds). The greater use of non-cabinet MPs on *The Daily Politics* was probably largely due to the frequency and timeslot of the programme, and specifically to availability issues for more 'prestigious' guests, rather than a preference for less powerful voices. Thus, easily the longest interview in *The Daily Politics* sample, at 13 minutes, was with Liberal Democrat leader Sir Menzies Campbell (18 May 2006). He was the most senior politician that appeared in the sample period. The public were best represented on *Question Time*, as studio members making up 19 per cent of speaking time on the programme by contributing both questions and (usually brief) comments. This was almost double the percentage for the next highest, *The Daily Politics* (11 per cent). Ordinary people accounted for less than 1 per cent of speaking time on either *The Politics Show, Sunday AM, Jonathan Dimbleby* or *This Week*, and 6 per cent on *Sunday Live*.

Discussion

Despite concerns over declining coverage of politics and other 'hard' news topics, we found fairly high levels of news about politics in our sample. As we discussed in Chapter 5, this varied from programme to programme. But overall, 'politics' was the most commonly reported domestic topic, with stories dealing with British politics (as a *'principal/ main'* subject) taking up one-third (31 per cent) of all news time and comprising nearly half (46 per cent) of all 'lead' or 'first headline' stories. Our analysis focused on national news bulletins rather than local news. It was perhaps predictable therefore, that much of the domestic political coverage would focus on national (i.e. Westminster) issues, events and personalities. But even accounting for this, the proportion of political news directed towards Westminster was striking (a staggering 97 per cent), especially given that our sample coincided with local government elections. Northern Irish, Scottish and Welsh affairs (3 per cent), along with local government issues (5 per cent), were seldom reported by comparison. Indeed, of the 27 stories about local government in our sample, only 4 did not give heavy focus to the Westminster scene. Rather, stories about local government, most of which related to the elections of 4 May 2006, were usually analysed in relation to Westminster events and personalities (e.g. 'Tomorrow voters in England will get the chance to elect local councils. It will be the biggest test of public opinion since the rows about prisoners, health service reforms and John Prescott' (*Newsnight*, BBC2 22.30, 3 May 2006)). As several local councillors were shown in the news to remark, the local election outcomes were decided primarily on the basis of the actions (especially mistakes) of Westminster MPs. The news media's focus on the national agenda during the campaign probably contributed to this situation reinforcing the sense that local elections function largely as a proxy for the national political scene. Our analysis further demonstrated that even within the Westminster-dominated news agenda, there was overwhelming focus on the three 'major' parties (Labour, Conservative, Liberal Democrats) and, within these parties, on a small 'elite' of senior ministers. In line with Bennett's 'indexing' hypothesis, we found that attention and media access tended to be 'indexed' to the proximity of political actors to the levers of power. Hence, Labour was a main story focus much more often than the Conservatives (fully 7 times as often by our measures), the Conservatives featured twice as frequently as the Liberal Democrats, and the Liberal Democrats appeared more often than any of the minor parties. As this reflects, most political news in our sample tended to revolve around the Labour government, with Conservatives and Liberal Democrats featuring as relative 'bit players', usually to provide a comment (typically a brief sound bite) on issues or problems concerning their Labour opponents. We should note that the greater prominence of Labour, and Prime Minister Tony Blair in particular, owed something to the international responsibilities of a government and Prime Minister. These responsibilities meant Labour and Blair often appeared in stories about foreign policy issues, while other politicians seldom did. Yet, even excluding foreign affairs stories, Labour politicians (n = 386 stories) were *'main'* foci in more than six times as many stories as Conservatives (n = 59) and more than 16 as many as Liberal Democrats (n = 23). Labour's prominence was also evident among news sources. Labour sources comprised two-thirds of all speaking time for 'political' news sources and spoke almost 4 times as much as Conservatives and 7 times as much as Liberal Democrats.

Politicians from any of the minor parties (e.g. Respect, Greens, BNP) contributed only 3 per cent of the total speaking time by political actors. Of the 'minor' parties, the BNP received the most attention following their surprise success in the local government elections, where they won 11 seats on London's Barking and Dagenham Council. Most of the attention was rather deflating however, usually raising alarm over the centrality of race in their policy stance (e.g. 'The British National Party ... has come under sustained attack from church leaders and political opponents' BBC1 22.00, 2 May 2006), 'They might be the pariahs of politics but the far right is also celebrating tonight after its best election showing in years' (*ITV News,* 22.30, 5 May 2006), 'The BNP pick up seats: a fringe protest? Or a threat to democracy?' (*Newsnight*, BBC2 22.30, 5 May 2006). The (fleeting) attention given to the BNP as a minor party contrasts with the virtual invisibility of the Respect Party, which had comparable success in the 2006 local elections. They also won 11 council seats including the Birmingham Sparkbrook ward, a Labour stronghold, where Salma Yaqoob won a crushing 55 per cent of the vote. Yet, of only four stories featuring Respect, all four related to comments by MP George Galloway that were framed by the media as 'controversial' (first expressing admiration for Fidel Castro while in Cuba (*Channel 4 News*, 25 May 2006) and then stating that the assassination of Tony Blair by terrorists would be of a 'different moral order' to suicide attacks against civilians (*Channel 4 News, Newsnight, BBC News,* 26 May 2006). The success of Respect as a party in the local elections, offering what might be thought of as something of an 'old Labour', broadly socialist political agenda, was completely ignored by the television news. Thus outside of New Labour, Conservatives and Liberal Democrats, political forces appear to be either invisible or made visible precisely in order to underline that the mainstream parties constitute the only sensible political discourse available, and that outside of the main parties, politics can only degenerate into extremism. Within each of the three main parties, the focus also tended to gravitate towards the most powerful figures. Cabinet MPs were identified as *'main'* story actors in three times as many stories as other (non-cabinet) MPs, MEPs, Lords and party activists. These other, less powerful party representatives did receive almost as much speaking time as cabinet officials; yet, very often their speech related to the actions of party elites. It was, for instance, common in our sample for backbench Labour MPs to be interviewed about Tony Blair's leadership (when should he step down?), about illegal immigration and prisoner deportation scandals engulfing the Home Secretary (first Charles Clarke, then John Reid), and about a series of scandals or gaffes involving the Deputy Prime

Minister, John Prescott. Our results, therefore, seem to confirm 'the growing journalistic preoccupation with the activities of government and senior politicians to the relative neglect of the back benches and the near wholesale exclusion of minority parties' (Franklin 1997: 236). Yet our analysis also demonstrated that it is not only minority parties, but the general public as well, who are excluded from playing any substantive role in setting the political agenda. The public contributed merely 9 per cent of speaking time for news sources in 'political' stories and received much shorter speaking opportunities on average in comparison to either politicians or 'experts'. This is in line with a study of the 2001 British election by Lewis, Inthorn and Wahl-Jorgensen (2005), which found that the public was rarely invited to articulate substantive opinions on policy issues and often contributed only briefly via vox pops (21–2). They found that the public were largely constructed as passive, emotional spectators or 'consumers', rather than citizens (ibid.: 49).The broadcasters' preoccupation with the personalities and mechanisms of power was also evident in the results of our analysis of different 'news frames', which showed that political stories emphasising 'horse race/strategy' (40 per cent) and 'competence/integrity' (41 per cent) were both more common than stories examining 'policy issues' (26 per cent). Stories containing the first two frames were generally about struggles for power and the performance of (mostly) those wielding power. Thus, the most common stories with a strong 'horse race/strategy' focus were reports about parties' (especially Labour) and their leaders' popularity in the lead-up to the local elections (e.g. 'The latest polls put the Conservatives ahead of Labour by anything from 2 to 9 per cent' (*ITV News*, 22.30, 2 May 2006)), discussion of Tony Blair's future following a poor showing for Labour in these elections (e.g. 'The tensions between Tony Blair and Gordon Brown are in sharp focus tonight after a big cabinet reshuffle and renewed talk of a power struggle' (*Ten O'Clock News*, BBC1 22.00, 5 May 2006)), and debate over the status or 'position' of embattled ministers such as Home Secretary Charles Clarke and Deputy Prime Minister John Prescott (e.g. 'The pressure piled on Charles Clarke today' (*Five News*, C5 17.30, 1 May 2006)). Similarly, the most reported 'competence/integrity' issues were the scandals involving the Home Secretary and the Deputy Prime Minister: the first concerning the Home Office's alleged mishandling of foreign prisoner deportations and illegal immigration, and the second an extra-marital affair by the Deputy Prime Minister, which was followed by sometimes intense scrutiny of his effectiveness in office. Both of the main 'scandal' issues were cases of media 'feeding frenzies'. The alarm over ministerial incompetence at the

Home Office could probably be viewed as an instance of the more legitimate media concern over ministerial conduct and performance. Reaction to the revelations ultimately cost the Home Secretary, Charles Clarke, his job. The attention given to John Prescott is less defensible, however. Much of it was another case of media scrutiny of the personal affairs of public figures (e.g. 'And talking of colourful personalities, in the past month or so we've found out a couple of things that John Prescott gets up to in his spare time' (*Five News*, C5 17.30, 29 May 2006)). Prescott remained a focus of attention after being stripped of his department but keeping his large salary and perks (e.g. 'The Prime Minister made a point of praising John Prescott's negotiating skills today. He has to find something to praise to justify the fact that Mr Prescott retains his salary, his official residences and his government car despite being stripped of his department' (*Newsnight*, BBC2 22.30, 8 May 2006)).

Whether the focus on competence/integrity includes the serious or the salacious, it does remain targeted overwhelmingly at the individual. While scholars such as Cappella and Jamieson (1996) have claimed that such coverage may cumulatively produce a corrosion of public confidence in 'the system' itself, it should be noted that this kind of scepticism, if it is derived mainly from media coverage, would be a very depoliticised one ('they are all as bad as each other') rather than one informed by an understanding of systemic problems and causal tendencies. While the focus on competence/integrity can produce major conflicts between political elites and media producers, it in no way signals interrogation or scepticism on the part of the media of major policy goals or principles. Stories dealing with such 'policy issues' comprised one in four of all the political stories in our sample (26 per cent). It is worth re-emphasising, however, that this frame was actually more prevalent than 'horse race/strategy' *after* the focus on the local elections of 4 May dissipated. The single biggest 'policy issue' in our sample was recurring focus on the Labour government's pensions reform package (e.g. 'If you're worried about your pension then there's some good news tonight: Tony Blair and Gordon Brown have struck a deal that will mean the basic state pension going up a lot faster' (*Five News*, 17.30, 12 May 2006); 'Women over the age of 45 are set to be the big winners in the most significant pensions shake up in 60 years' (*Sky News*, 22.00, 18 May 2006)). This frame is widely seen as an instance of substantive political coverage, yet it cannot be regarded as unproblematic either. As the first of these headlines (concerning the 'deal' struck between supposed rivals Blair and Brown) suggests, focus on policy can often be permeated with horse race or contest framing. But perhaps even more crucially, the slight increase in 'policy issue' framing once the election focus had died down is more than tempered by the evidence of an overall domination of policy debates by a narrow hierarchy of political elites. Hence it may well denote a focus on issues within an extremely narrow range of debate, especially given the political consensus that has been forged across the mainstream parties on many issues. Taking the pensions issue as an example, no mainstream party questioned the adequacy of private pension provision or the low rate of contribution to the pension pot paid by the private sector. This was in turn reflected in much of the news reportage which framed the deal as 'good news' (see *Five News* quote above). According to Labour MP Austin Mitchell, writing outside the mainstream media:

> Successive British governments have been too afraid to upset corporations by requiring them to make adequate national insurance contributions. British employers are typically required to pay 12.8 percent of employee earnings ... This contribution by British employers translates as approximately 9.6 percent of their labour costs, as compared to an average 15.2 percent for OECD countries and an average 17.8 percent across the European Union (EU). In many EU countries the proportion is much higher – for France, Italy, Belgium and Austria, the figures are 29.7 percent, 24.9 percent, 23.3 percent and 22.6 percent respectively. (Mitchell and Sikka 2006)

Many of the findings of our news sample were also evident in our analysis of political discussion programmes. With their normally longer duration and their potential for a less event-driven focus, these programmes are seemingly better equipped to tackle 'policy issues' than standard news bulletins. The additional hours of political or public affairs programming they provide also opens up the opportunity for a more diverse range of voices to contribute to political discourse than is provided on the news. Dealing with the question of diversity first, we found that there was little variation between the choice of sources on political discussion programmes compared with the news. The proportion of speaking time given to politicians was lower than on news programmes (56 per cent compared with 66 per cent), but politicians still clearly dominated the debate. Likewise the levels of speaking time for 'experts' and the public ('ordinary' people) were surprisingly similar across the two formats. With the exception of *Question Time*, on which ordinary people contributed about one-fifth of speaking time, the public was again largely excluded from the discussions (receiving only

8 per cent of speaking time). The question of 'framing' is more difficult to adjudge, especially given that marked variations in format mitigated against reliable quantitative comparisons. What became obvious upon analysing these programmes, however, was that the issues they discussed only occasionally diverted from the mainstream news agenda. While these discussion programmes did sometimes investigate an issue that had been largely overlooked on the news, for the most part they focused on the same issues but in more depth. This may not adequately address the under-representation of certain important issues on the news, but it has the advantage of enabling, at least potentially, a more detailed exploration of key policy issues. And we did find that this often happened. But we also found that many of the events and issues that were discussed in the most detail had only marginal 'policy' relevance. Hence there was a great deal of discussion, particularly in the first fortnight of our month-long sample, about 'horse race/strategy' topics (such as Labour's waning popularity, the local election results, the effectiveness of David Cameron and Menzies Campbell as party leaders, the 'power struggle' between Tony Blair and Gordon Brown, a Labour cabinet reshuffle on 5 May), and 'competence/integrity' topics (such as the Home Office 'fiascos' and John Prescott's troubles). As with the news sample, discussion of policy issues did increase in the latter half of the coding period, however. Particularly in this latter phase the potential for these programmes to provide more detailed policy debate than the news became more evident. Among the policy issues discussed in much detail across the seven political discussion programmes were the war in Iraq, immigration, pensions reform, education reform and Britain's Human Rights Act. The education reforms in particular received only scant coverage in our news sample and this coverage was generally light on policy substance. In contrast, coverage of this issue was at times significantly more detailed on the discussion programmes, including some particularly long analyses on *The Daily Politics* (9 minutes on 23 May, 13 minutes 24 May) and *The Politics Show* (14 minutes, 21 May). Similarly, news about Britain's Human Rights Act was often short and focused upon sensational cases such as a court ruling against government plans to deport a group of Afghan asylum seekers who hijacked a plane to Britain. On the discussion programmes the pros and cons of this Act were debated at length, including 23 minutes on *Question Time* (across 11 and 18 May), 20 minutes on *Sunday Live* (14 May), 13 minutes on *The Daily Politics* (15 May), 8 minutes on *Sunday AM* (14 May) and 7 minutes on *Jonathan Dimbleby* (7 May). However, as the results for the sources for quotations on the political programmes suggest, the problem

of limited political or ideological diversity was by no means overcome. In the case of the Human Rights Act, for instance, the discussion was overwhelmingly critical of the legislation and there was little clear dissent to calls for it to be significantly reformed or even repealed. There were some instances of alternative or 'outer-consensus' voices expressing critical opinions, such as interviews with entertainers Stephen Fry (*This Week*, 18 May 2006) and Terry Jones (*Sunday Live*, 14 May 2006) and animal rights activist Keith Mann (*Sunday Live*, 14 May 2006). But debate was still clearly controlled by politicians from the main parties, New Labour and the Conservatives in particular.

Conclusion

At a time when public attitudes and behaviour and academic studies and reports are calling into question the quality and effectiveness of our major institutions of political representation, the news media's interlocking relations with those institutions is suffocating debates about policy directions that might genuinely be based on public consensus through participation and dialogue. Instead, the television news media have constructed a graduated hierarchy of access and focus that assumes the representativeness of political elites and excludes public voices or alternative political programmes (except, in the case of the BNP, to imply that beyond the mainstream parties there can only be extremism). Thus in a rigidly structured hierarchy, the Prime Minister dominates over the cabinet, the cabinet dominates over ordinary MPs, the governing party dominates over the opposition, the three main parties dominate overwhelmingly over smaller parties, and the political elites dominate over ordinary members of the public. This hierarchical picture is very different to that painted by liberal media scholars, where diversity and plurality and public participation appear to characterise contemporary broadcasting. Our findings uncover a remarkable skew towards the governing party. We are not suggesting a pro-Labour bias (least of all because much of the coverage was problematic for the Labour party) but instead a bias towards whoever occupies the formal positions of political power. The attraction of the media towards those occupying formal positions of power also explains the skews relating to cabinet ministers over non-cabinet ministers. No doubt the media would justify such skews in terms of the assumed representativeness of elected Parliamentarians. But this glosses over the widening disquiet over this assumption. First, there is the long-standing concern over the UK's electoral mechanism (the UK's winner takes all/first past the post system for national elections). This concern has led to alternative voting systems for regional assemblies in Northern Ireland, Scotland and Wales as well as for the London Mayor (Miller, Timpson and Lessnoff 1996: 189–204). However, as we have seen, for the national news, politics is synonymous with Westminster. Regional politics apparently has little national significance. And beyond the concerns about electoral mechanisms, there is the broader concern about the subordination of policy-making to market imperatives and corporate actors. We have found that continuing hierarchical structures are coupled with media coverage of politics in which political discourse is largely reduced to issues of the 'horse race' and anti-systemic enquiry into competence and integrity. We found that while 'horse race' coverage dominated massively during the election campaign, competence issues focusing on Home Office Minister Charles Clarke and integrity issues focusing on Deputy Prime Minister John Prescott, dominated thereafter. The terms of criticism by media scholars which have been levelled against such trends are often inadequate, failing to understand the lines of complicity as well as tensions between political elites and media elites, as well as the relations of both to wider socio-economic forces. In other words we must grasp the intricate connection between the hierarchical structure to political reportage and the narrowly conceived populism through which the media offers a critique of politics. Although policy issues do make a notable comeback outside the campaign period, we cannot even see this category as an unalloyed 'good', given the deeply hierarchical and exclusive way the political debate is structured. This Westminster-centric monopoly on political discourse and debate is likely only to exacerbate public scepticism towards both the news media and mainstream politics. Like politics itself, media coverage of politics is an elite and exclusive affair, a spectacle for the public to consume rather than a process for them to enter in order to democratically discover what might be the public good on any given issue.

7

The Boundaries of Political Debate: Animal Rights

According to the UK's Office of Communications (OFCOM), television news 'is important because it informs and educates citizens, helping them take part in the democratic process' (OFCOM 2007c: 12). There are certainly good reasons why aspects of this proposition should be a *norm* to aspire to, but whether it is substantively a *fact* in the here and now is another question. As a norm to aspire to, effective participation in the democratic process (whether that is in periodic formal electoral politics and/or various other activities of persuasion and campaigning) requires knowledge, information and understanding of the issues involved and a grasp of the strengths and weaknesses of opposing arguments. The media in all its forms and at different scales of operation obviously play an important role in the quality and complexity of that knowledge, information and understanding. But we must suspend OFCOM's assumption that television news (still one of the most important sources of news for the public) does this job of informing and educating effectively (from the point of view of adequately informing consent and/or dissent with this or that aspect of the social process). We must instead interrogate it. Our study unusually observed not just the daily television news bulletins but also the range of political discussion programmes that were available during the same period. This allows us to explore the interrelationships, overlaps and differences between the daily news bulletins and the political discussion programmes in relation to specific issues that flared up during May 2006. Our aim in this chapter is to explore the extent to which television news *and* the broader range of political discussion and information programmes can be said to offer the public knowledge, debate and an exchange of ideas on the substance of an issue that has public policy and private/corporate implications. Our case study focuses on the animal rights protests or, to be more accurate, the anti-vivisection protests and campaigns that were in the news throughout May 2006. There were 38 stories about animal rights on the 6 news programmes that we analysed. The breakdown was as follows: BBC *Ten O'Clock News* (5), BBC2's *Newsnight* (5), *ITV News* (3), *Channel 4 News* (10), *Five News* (9) and *Sky News at Ten* (6). That amounts to 2 per cent of all stories analysed (2304). They took up 103 minutes of news time or 2 per cent of all story time. Beyond these 'news' programmes there were a number of other programmes that examined the question of animal testing. These were Sky News' *Sunday Live*, BBC1's *Question Time*, Channel Five's *The Wright Stuff*, a weekday morning topical discussion programme, and a Channel 4 *Dispatches* documentary about animal rights activists that was broadcast in the middle of the month (15 May 2006).

The media and social protest

Indications of how television news (and the wider media) might treat the animal rights campaigns of May 2006 can be gleaned from previous studies into media representation of protests and direct action campaigns. We can define protest politics as politics that takes place outside the sphere of formal electoral politics and its institutions (McLeod and Detenber 1999). Despite the growing evidence of a crisis of legitimacy within these formal democratic institutions, many studies of the media demonstrate that they continue to see the politics of protest as intrinsically deficient, illegitimate and less representative of broader public opinion than the representatives of formal electoral politics. This is the case even when over the course of time, a new consensus has established itself in some alignment with precisely the concerns and issues which social protest had been trying to raise. Todd Gitlin's classic account of how the American media dealt with both the civil rights struggles for black people in America in the 1960s and the anti-Vietnam war protests is a case in point. Few would now dispute that the lack of black civil rights in America was a human rights scandal, and it would be uncontentious now to suggest that America's involvement in Vietnam was detrimental to both America and Vietnam. The media coverage of both these social movements, however, was one that was wedded to the very institutions responsible for discrimination and war. As Gitlin notes, social movements and social protest acquire a presence in the mass public sphere, that is, become 'news', only by simultaneously becoming very largely defined and made meaningful within particular parameters or frames that belong overwhelmingly to the media rather than the movements themselves (Gitlin 1980: 3). These frames are

> persistent patterns of cognition, interpretation, and presentation, of selection, emphasis, and exclusion, by which symbol-handlers routinely organize discourse, whether verbal or visual. (Ibid.: 7)

Some theorists have argued that the notion of news frames has recently become separated in media analysis from the question of the power relations in which 'symbol-handlers' are immersed and often uncritically aligned with (Caragee and Roefs 2004). Others, such as Cottle and Rai (2006) complain that the concept of 'frames' has conversely been too securely stapled to the notion of a 'preferred point of view' or 'dominant discourse'. They suggest that this 'too easily slides over the formal complexities of the communicative architecture of television news' (170). They cite television reception studies which find audiences 'are in fact complexly involved in reading processes' (170) that demonstrate evaluation of both the topics of news and the manner of its presentation. But this is to conflate two distinct propositions. To say that news frames tend to privilege certain perspectives most congruent with political and economic power does not imply that audiences are passively lapping up such points of view. But it is to say that audiences have to work with the materials they are presented with and thus may not gain much access to other potential materials which might be better suited to expanding audiences' critical engagement with the issues at hand. Moreover, there are also plenty of studies which indicate that complexity of audience responses notwithstanding, television genres, including news, *do* have some determinate effects on many people at least some of the time (see Behr and Iyengar 1985; Dixon and Azocar 2007; McLeod and Detenber 1999; Nelson, Clawson and Oxley 1997; Philo 1990). While there is certainly some diversity in the 'communicative architecture' of television news (that is the different forms and strategies of presentation it has at its disposal), the key issue is to what extent this diversity correlates with a genuine diversity of perspectives and cognitive paradigms outside the dominant institutions in any social order. One of the most subtle and effective media frames is that 'protest' constitutes a disruption to what is otherwise a legitimate and well-managed social order (Gitlin 1980: 271). Business as usual (as long as laws do not appear to have been contravened) is very rarely represented as a disruption in the moral, social or political fabric. This presumption tilts the media terrain decisively in favour of powerful actors who have access to the media from positions of state and corporate power. Bennett et al. (2004: 438) suggest that the quality of debate within public spheres can be assessed by looking at three characteristics: access to the media (who has it), recognition (how are speakers identified and to what extent there is differential legitimation between speakers), and responsiveness (how much dialogue there is between protagonists in the debate and are their different claims put to each other?). Their case study is of the influential *New York Times'* representation of the World Economic Forum (WEF) and the World Social Forum (WSF) between 2001–03. The WEF is an international organisation that plays host to major meetings of political and business leaders. It has a leading role in advocating international capitalist competition and globalisation. In contrast, the WSF is a gathering of social movements and networks contesting various aspects of capitalist globalisation (such as privatisation, deregulation, the downgrading of labour rights, war, debt in the developing world and so forth). The Bennett et al. study found that access to *The New York Times* was overwhelmingly dominated by WEF advocates, that WSF participants were delegitimised in the paper and that the 'activist' frames (such as concerns about international debt) that appeared in the media and WEF discourse as a result of WSF public pressure were appropriated and made congruent with the dominant priorities of capitalist profitability and competitiveness. In other words there was limited responsiveness:

> Instead of taking the claims of each side to the other for reaction, the press framed the protesters in ways that limited their legitimacy, marginalized their status, and ultimately gave ownership of their issues to the higher status actors on terms those actors chose. (Ibid.: 452)

As protest and direct action often involves at the very least occupation of public spaces that might be contested between protesters and the state, the media also routinely use a range of 'marginalization devices' (Dardis 2006) to undermine or divert media focus away from the beliefs of protesters and the causes which have pushed them into political action. These include focusing on clashes with the police, instances of law-breaking, disrupting or disturbing the activities of the 'normal' citizen and the casting of protesters as, in various ways, 'juvenile'. Even the protests against the American/UK attack on Iraq in 2003 were treated as less legitimate public action than state sanctioned violence that horrified majority public opinion the world over and which even at the time was regarded by many people as having questionable legal basis in international law. Once again, the consensus view *now* is that the war on Iraq and its aftermath has been an unmitigated disaster for both the West and the people of Iraq. But in both the run up to the war and during its prosecution, the media failed to critically scrutinise the arguments of the pro-war case and instead became champions of it (Edwards and Cromwell 2006; Miller 2003). This alignment between the media and the state necessarily meant that protests against the war, especially once it had started, had to be treated in such a way as to undermine them rather than engage substantively with the points the protesters were making. Cushing found that while UK coverage of young people's involvement in protest was open to framing such activities as legitimate, once the war started, the media swung behind state interests and the coverage of anti-war protest shifted accordingly. Young people of school age were involved in a number of protests shortly after the attack on Iraq started, which involved 'walk-outs' from school. Now the media (which had a few years before been obsessing about a supposedly apathetic and apolitical youth) focused on the issue of 'truancy', disruption and 'law and order' issues to delegitimise the protests. As Cushion notes

> Press coverage was therefore not framed around young people taking principled, politically motivated, citizenship stances, but by blatant opportunism; of using politics as an excuse to escape from the responsibilities of being young. (Cushion 2007: 428)

Driving the issue: Debate or powerful interests?

Animal rights, as a cause around which direct action could be mobilised, must count as an example of the proliferating political movements of civil society that has been growing over the past 30 or 40 years. In the UK, the 1990s saw growing campaigns against road building, criminalisation of youth activities, third world debt and worker exploitation: all emblematic of the new modes of organising and new sites of conflict outside the classic confrontation between capital (and the state) and labour at the point of production. In 1995 protests exploded onto the national scene at Shoreham and Brightlingsea over the export of veal calves. Benton and Redfearn found that while the media coverage of these new protests employed some familiar strategies (such as focusing on 'violence' from the protesters, but not the police), the coverage was 'far removed from the hegemonic hostility which has characterized media treatment of industrial action by workers ...' (1996: 44). The issue in May 2006, however, was specifically about vivisection or experiments on animals for the purpose of medical science. As we have seen in relation to the anti-Iraq war protests, media coverage can shift in tone and approach once the media perceive that the fundamental interests and legitimacy of state and/or capitalism are at stake. It is this sensitivity to powerful interests rather than the need to construct a debate for the public that was the main driver in the coverage of the anti-vivisection campaigns throughout the month. Crucially, two of the stories involved a nexus of very powerful interests that were involved in a carefully coordinated counter-attack against the animal rights/anti-vivisection protesters: these included the government (especially Prime Minister Tony Blair), the state more broadly (especially the police and the courts), the pharmaceutical industry (especially GlaxoSmithKline) and Oxford University which was in the process of building a £20 million animal research laboratory. The outcomes of protest campaigns are crucially linked to the responses of the organisations under attack. Jasper and Poulsen have argued that the early successes of protest movements in turn provoke state repression and counter-mobilisation by targeted groups, which in turns leads to decreasing success over time (1993: 640). In America, the level of counter-mobilisation by targeted groups involved in animal vivisection increased over time, while success by anti-vivisection campaigners decreased (Jasper and Poulsen 1993: 646). The role of the media generally and television specifically in May 2006 certainly seemed to be to work as a fairly uncritical conduit of a powerful counter-mobilisation by a nexus of interests. This meant that the 'winning or losing' of public support (for or against vivisection), rather than an engagement with the substantive issues and the nuances of the debate, was the main focus and effect of the media. The public were asked to choose between an animal rights position on the one hand and a business as usual position from the pharmaceutical industry and government, while at the same time, the latter were using the issue of the *tactics* of animal rights protesters (the main focus of the media) as a reason for increasing state powers over the right to protest. The large middle ground between the position that animals have 'rights' and the business as usual position adopted by government and industry remained unexplored. If the notion that animals have rights is philosophically incoherent (animals cannot articulate their rights, nor extend such a notion to other animals within or beyond their species) the notion that human beings have duties and obligations towards them and that how we treat animals is a proper subject of moral concern has a very wide degree of consensus within Western society (Benton and Redfearn 1996: 51; Rose 1992: 150). A number of issues within these parameters *might* have been expected to have cropped up in the course of the month, if the television news and political discussion programmes could claim to be functioning as OFCOM hopes: that is to inform and educate citizens so that they can influence public policy from an informed position. For example, to what extent are there alternatives to animal testing for medical research? How confident can the public be that all procedures are actually necessary and potentially beneficial to science? How rigorous is the monitoring procedure within the research laboratories? This suggests that we should add a fourth criteria to the three which the Bennett et al. (2004) study used for assessing the quality of the media's contribution to the public sphere. To access, legitimacy and responsiveness, we must ask for the media themselves to exercise their capacity for autonomous investigation and critical editorial judgement *when the discursive positions of the protagonists in a debate fail to exhaust all the areas of public interest that might be at play.* Arguably this is the case here in the 'middle ground' between the animal rights position and the business as usual position of industry and government.

Digging up the body, burying the issues

The issue of the tactics of the animal rights protesters was the overwhelming focus of the television news media (ably supported by the national newspapers) with a heavy emphasis on law-breaking and on threatening behaviour and intimidation. In terms of Bennett et al.'s criteria for assessing the quality of debate within the public sphere, this focus within the media underpinned very unequal access to the debate, strongly differential legitimation between participants, and with one or two exceptions, generally limited responsiveness (so that the claims of each side are generally not responded to by the other side). On 3 May 2006, the BBC news ran the story about Gladys Hammond's body which had been removed from her grave by animal rights protesters a year before. Gladys Hammond had been a relative of the Hall family who owned a farm that bred guinea pigs for animal experimentation and which had been targeted by animal rights activists during a long campaign. As the trial of four activists said to be involved in the offence was coming to a close, the police were searching for the body in a forest. The news item also covered and connected this story with the protests at the animal research laboratory under construction in Oxford. The Gladys Hammond story and the animal research laboratory under construction in Oxford were to run throughout the month and were joined by a third story on 9 May. This concerned a letter that had been written to a number of small shareholders in GlaxoSmithKline (GSK), a British pharmaceutical company and one of the largest in Europe. They were asked to sell their shares because of GSK's links with the animal research laboratory Huntingdon Life Sciences against which animal rights protesters had long campaigned. The letter stated that should shareholders not sell their shares, their names and addresses would be published on the Internet. The BBC reporter, *business* editor Robert Preston (9 May) described this as a 'sinister warning'. Sky News described the letters as part of an 'animal rights hate campaign' (9 May). Back on the BBC, an animal rights protest march was shown in slow motion, in black and white and with a focus on figures in balaclavas. In contrast, in the same report a little later, a march by the Oxford-based pro-vivisection group Pro-Test was shown without any special effects rendering them 'sinister'. That the images used to illustrate news stories are important can be gauged by the 9 May edition of *Newsnight* which featured both a report into the letters to shareholders and a subsequent studio debate to which we will return. *Newsnight* anchorman Jeremy Paxman introduced the report with a photograph on the screen behind him. While verbally the report was about the letter writing campaign and the 'militants' behind it, visually the photograph was a reminder of other forms of protest by people whose appearance does not connote the same sense of threat that men in balaclavas do. Here the image was a close-up of a protest placard with the words 'Laboratory Animals Need Your Help'. There is a picture of a dog on the placard with the words 'Born To Die' above it. The placard is held by two blonde-haired women, one middle aged, one older. This is an example of how the multiplicity of signs in news coverage do not necessarily cohere into a seamless, univocal meaning. The image provides some semiotic disruption to the tendency of the news coverage to tar all protest with the same brush. But as we shall see, over the course of the month, the negative focus on the letter campaign and the Gladys Hammond story *did* begin to raise questions about whether and to what extent a whole range of protest activity may or may not be legitimate. To what extent was the media response to the letter writing campaign justified? As the information on the small shareholders was already publically available via Companies House, it was not entirely clear what, if anything, was illegal about the letters. Channel 4 (9 May) did at least explain that the letters could be illegal if the wider dissemination of this information could be said to constitute some sort of threat. Where animal rights campaigners were invited to speak, it was solely in terms of whether they would condone or condemn this or other *tactics*. Thus Poorva Joshipura from People For The Ethical Treatment of Animals (PETA) was introduced on the Channel 4 item by the reporter with these words: 'Mainstream animal rights activists refused to condemn it [the letter writing campaign]'. *More Four News* meanwhile, with its more youth orientated focus, flew 16-year-old Laurie Pycroft, one of the founders of Pro-Test, to America for a interview with animal rights activist, Nicholas Atwood. But the 'debate' such as it was focused solely on the legitimacy of direct action tactics rather than the substantive claims underpinning animal rights action or conversely the *moral case for* animal experiments for medical research. The report on *Newsnight* on 9 May also featured PETA (again being asked about tactics). And there were four other interviewees all in some way protesting against animal rights activists. The first is a Dr Nick Steiner, one of the small shareholders who received the letter. He describes the letter writing campaign as a 'terrorist act' because, he argues, it is seeking to destroy a company. The linkage between animal rights tactics and terrorism was to become a common theme in the month-long debate and we shall return to it later. The second interviewee was an 'undercover reporter' (here we see only his shadow) who is strongly condemnatory of the tactics of the Animal Liberation Front which he has infiltrated. The third interviewee is a representative from the UK Shareholders Association advocating tighter control over the release of commercial information. Then comes Poorva Joshipura again from PETA and finally a spokesperson for the Biomedical Research Organisation who criticises the New York Stock Exchange for pulling out of listing the American subsidiary of Huntington Life Sciences on the exchange after pressure from animal rights protesters. It is clear that within this report there is unequal access to the debate, differential legitimation and no exchange of claims between them. For example, no one on the animal rights side of the argument is asked whether the description of the letter writing campaign as an act of terrorism is proportionate and accurate. It is interesting that the mainstream animal rights movement was represented by PETA on so many news reports. While PETA is well known in America where it began, it is less well known to the British public than an organisation (and a very *establishment* organisation at that) such as the Royal Society For The Prevention Of Cruelty To Animals (RSPCA). Yet the RSPCA did not feature in the television news reports. This is not because the RSPCA has no position on the issue. A letter in *The Independent* (20 May) from Jackie Ballard, Director General of the RSPCA argued that many experiments 'are carried out on animals for a broad range of purposes, many of which have little or no direct medical application. Industry, government and the scientific community should give far higher priority to replacing animal experiments.'[1] There can be little doubt that the absence of the highly respected RSPCA from the television screens during the May 'debate' on vivisection helped further to construct the 'business as usual' position which the research community, business, government and the dominant media were committed to.

Newsnight and differential legitimacy

Following the report into the letter campaign against GlaxoSmithKline, *Newsnight* (9 May) had a debate between 19-year-old Iain Simpson, one of the founders of the Oxford pro-vivisection organisation Pro-Test, and John Curtin, an animal rights campaigner. Curtin joined the debate from an exterior location somewhere in Birmingham while Iain Simpson joins Jeremy Paxman in the *Newsnight* studio. We cannot know what logistical reasons lay behind John Curtin's exterior location but the signifying (and practical) effect of the contrast between his location and Iain Simpson's (within the inner sanctum of the studio) becomes a major contribution to the differential legitimation which the programme gives the two speakers.

Simpson's presence within the studio is significant in the wider context of this study. We have seen that young people contributed only 3 per cent of speaking time by news sources across the sample news programmes (which included *Newsnight*) for the month of May 2006. Half of these speaking opportunities focused on the issues of crime and sport, which were the subjects in relation to which young people figured most. The studio interview with Simpson was one of only two throughout the month-long sample in which a young person (Simpson was 19) was invited into a television news studio setting to discuss any topic (the other was a *Channel 4 News* interview with another Oxford student, speaking about AIDS in Africa, 9 May). Interviewees invited into a studio setting tend to be overwhelmingly politicians and professional experts. Because the studio is the site where the news anchor or presenter is located, it is by association a high-status zone and an interviewee's presence within it affords the interviewee a certain assumed authority to speak on the issues at hand. That Simpson, a white, male, middle class Oxbridge type defending the status quo establishment position within the debate is one of the only young persons to make it into the inner sanctum of the studio within our month-long sample tells us something about the values and orientations of the media towards young people and politics. (The other young person was black and female, and while this is to be welcomed, the topic was 'international' rather than about the domestic/national political agenda).

These values and orientations are strongly shaped by class background. The visual division between John Curtin on the screen, outside (and with his working-class voice) and Jeremy Paxman and Iain Simpson *within* the studio is strongly underscored by the Oxbridge alliance between Paxman and Simpson. One give-a-way clue in the class sympathies of Paxman comes early in the interview when Paxman interrupts Curtin (who during the interview described the earlier report into the letter writing campaign as 'hysterical' and compared it with the news values of the *Daily Mail*) to say: 'hang on a second matey'. Would he have described a 'professional' or a politician or indeed Iain Simpson in these terms? Unconsciously no doubt, Paxman's language indicates that he is well aware of the class differences at play between his guests and interlocutors. Following the news agenda established by the other bulletins, Paxman focuses first on the question of the tactics of the letter writing campaign (although Curtin has no connection with that campaign). Paxman does put to Simpson Curtin's point that this information is already in the public field and presses him to explain why this campaign might be regarded as intimidatory. So here there is some sense of responsiveness within the debate. However, after this, Paxman's questions to Simpson can be best described as invitations to elaborate on the situation rather than interrogations of his position. For example, Paxman subsequently invites Simpson to talk about the growing backlash against the animal rights protesters. In contrast, Curtin's attempts to shift the debate onto terms that might raise questions about the unimpeachable goodness of the institutions involved are dismissed by Paxman. Curtin cites a case where GSK were involved in drug trials on orphans in New York and he advises viewers to google GSK and 'orphans' as a follow up to the programme. Paxman, however, dismisses the accusation as 'tittle-tattle' and is even aided by Simpson who begins to occupy an almost co-role with Paxman by saying 'can we move on'. Paxman replies: 'ok'. In fact the GSK case cited by Curtin came to light in 2004, and was the subject of a BBC2 documentary. That Iain Simpson is well within his comfort-zone through out the interview is indicated by his body language (he is even talking with his hands behind the back of his head at one point) and his occasional inaudible comments to Paxman coming off-screen when Paxman is talking to Curtin. Visually, the strong ideological bonds between Paxman and Simpson is nicely illustrated in an extraordinary moment where both Simpson and Paxman (in shot) raise their arms up to Curtin simultaneously in a gesture of 'be quiet' as he suggests that the current debate is too limited and there needs to be a Royal Commission into the question of vivisection. 'Yes' replies Paxman, 'we've heard it.' Paxman's response is indicative of the lack of legitimacy which television news accords to sources to comment *reflexively* on the framing conditions under which they enter the mediated public sphere.

Illustration 7.1 *Newsnight's* differential legitimation
Source: Courtesy of *Newsnight*.

Is it terrorism?

On 11 May, the television news returned to the story of Gladys Hammond and the animal rights intimidation of the guinea pig breeding farm, when four activists were jailed for their part in the campaign. Three men were sentenced to 12 years and one woman was sentenced to 4 years. On *Channel 4 News*, anchor Jon Snow introduced the story on the sentences thus: 'it is being held as a major blow against another form of domestic terrorism'. As we have seen, the framing of animal rights activists' tactics in terms of terrorism was a significant feature of media coverage. Following the 9/11 attacks on America, the phrase 'war on terror' was coined by the Bush White House and uncritically recycled by television news networks (Lipscultz 2007). The discourse on the 'war on terror' aims to draw a self-righteous line between who is 'fighting terrorism' and who is 'waging terrorism'. It has been an extremely useful linguistic and ideological weapon within American politics designed to silence dissent and negatively label opponents. The aim is evidently to 'produce patriotic, docile subjects' (Puar and Rai 2002: 130). This in turn requires a certain elasticity in the application of who is or is not a terrorist, which can only be sustained by the absence of critical scrutiny in the mass media. During May 2006, the media closely linked the framing of animal rights tactics as 'terrorism' with the pronouncements of senior politicians. A headline in the national newspaper *The Mirror*, 'Blair: No To Animal Terrorism' (15 May) was typical. Irrespective of whether Blair actually used the word terrorism in public, this authorises an equivalence between what many would regard, in other instances, as vandalism or threats, with the kind of spectacularly bloody and fatal violence on a mass scale, usually associated with terrorism. By the end of the month *The Daily Telegraph* could report that a YouGov survey for the paper found that 'more than three-quarters believe that the more fanatical activists can justifiably be defined as "terrorists"'.[2] Yet, in order for such a consensus to rest on rational foundations, it ought to have emerged out of some debate and reflection on whether this was an appropriate application of the term terrorism. But no such debate took place. Instead the term 'terrorism' was used but never scrutinised. The level of orchestration between the media, senior political figures and pro-vivisection groups within civil society was illustrated over the weekend of 13–14 May. On 13 May, Sky News (owned by Rupert Murdoch) ran the story that Tony Blair had written a newspaper article for *The Sunday Telegraph* stating that he would be shortly signing an online petition campaigning for animal testing. Yet the article was coming out the *following* day in a paper that was not part of the Murdoch empire. This indicates that behind the scenes, there was considerable orchestration going on between the Prime Minister's office and the media. An article in *The Sunday Telegraph* on 14 May would seem to be correct then when it stated

> The one bit of good news to emerge from last week's unprecedented attacks is that by targeting a company as big and valuable to the City as GSK, the animal rights extremists have finally gone a step too far. Investors might have no qualms about deserting a business such as Huntingdon, but they cannot afford to see one of Britain's largest and most valuable companies come under attack.[3]

Sunday Live

It was not until 14 May that something like a debate which has the characteristics of a properly functioning public sphere (equality of access, no differential legitimation and responsiveness) emerged. The occasion was Sky's *Sunday Live* with Adam Bolton. After a very brief introduction by Bolton recapping the previous week's events (the letters to GSK and the Gladys Hammond story) there followed a nine-minute interview and exchange between Iain Simpson (introduced as the 19-year-old who helped set up Pro-Test) and the 'veteran' animal rights protestor Keith Mann. Unlike the *Newsnight* interview which had begun and focused much of the time on the question of animal rights tactics (peculiar when John Curtin had no involvement in the letter writing campaign), this interview began with Bolton questioning Mann about the moral basis of animal rights protests. Bolton asked whether Mann believed that animals are equivalent to human beings. Mann was provided with the space – for the first time really since the beginning of the month – to put a central plank in the animal rights case: that animals and human beings are *not* biologically equivalent to the extent that experiments on animals can help the safe passage of medicines from the laboratory to the drug stores. Bolton responded by carefully noting that this does not address the *moral* basis for animal rights action, because 'if you didn't care about animals, it wouldn't be a problem'. Mann made a brief reference to animals being 'living, sentient beings' but his discursive strategy was primarily aimed at linking the issues to human rights. There was then some evasiveness from Mann in terms of specifically addressing the moral basis for action as it concerned *animals* and no doubt this was a defensive strategy which anticipated a lack of public hearing for the notion of *moral* equivalence between animals and humans. The nearest Mann got to addressing this issue was where he noted that he would not necessarily say one human being is worth the same as a guinea pig, but then neither would he say that human being *x* was of equal moral worth as human being *y* – which was at least an interesting point, even if it again did not directly address the issue of where in the chain of priorities animals come. Bolton then put the argument about a lack of *biological* equivalence between animals and human beings to Iain Simpson, thus demonstrating the media as a realm where protagonists can respond to the claims of others. This is quite routine when the protagonists come from the 'establishment' (the political class, the business class, professionals and experts) but as we have seen was noticeably absent when it came to dealing with political activities outside the sphere of formal politics. Simpson responded to the question of biological non-equivalence between animals and humans by listing a range of drugs that have been developed from animal research – although perhaps a more substantive response would have been to explain why animal biology is in some ways coterminous with human biology in ways that cannot be replicated without animals (computer modelling, for example) rather than listing the results of medical research. We have seen that in the *Newsnight* interview, there was little or no discursive space for John Curtin to raise questions about the broader activities of the pharmaceutical industry. Here, in contrast, Mann repeated several times that 'the fourth biggest killer in the western world are products we've been given by our GPs that have been passed as safe by these drug companies [after tests] on animals in laboratories.' Perhaps because Mann did not name a particular company (and thus raise fears over lawsuits, as in the case of GSK being named on *Newsnight*) Bolton, unlike Paxman, did not dismiss this claim as 'tittle tattle'. Instead, he asked for an example of a drug from Mann and then put this example to Simpson. A debate ensued as to whether this example of a drug that had killed people invalidated medical animal research, with Mann suggesting that the animals died for no reason, and Simpson pointing out that the drug passed tests on human beings as well, but that a lot more human beings would die from faulty drugs without the tests on animals. Mann, like Curtin, called into question the limited nature of the public debate and indeed the limited nature of the very interview he was taking part in, to discuss the issue. He suggested that instead of having himself, which he ironically labels 'an animal rights extremist' (thus critiquing the media classifications of protesters) and Simpson ('I don't know who he is!'), there should be a debate by scientists on both sides of the issue. Again, reflecting critically on the media constructed boundaries of the debate, Mann invited Sky television to set up such a debate, because, as he put it, 'no one else is interested in doing it'. It was interesting that while on *Newsnight* Simpson claimed to welcome debate, here he was much less enthusiastic – instead arguing that such a debate had already happened, and telling Mann that he was 'talking rubbish' when Mann, in contrast, suggested that such a debate had not really happened (presumably within the wider public realm). In contrast with the *Newsnight* debate, the lack of a class-based mutual identification/alliance between Simpson and Bolton meant that the programme accorded equal legitimation to both parties. This might explain why Simpson was noticeably bad tempered during the exchange ('excuse me I'm talking!' he almost shouted at one point, while resorting to sarcasm elsewhere – 'you're made up statistics are very interesting'). The other issue that emerged from the debate was a question mark over whether animal experimentation for the purposes of cosmetic research had actually ended. Simpson cited the legal position that it had (since 1998) but Mann claimed to have broken into a laboratory two years previously and found rabbits being injected with botox. He also raised questions about the endless replication of experiments. This broadened the issue out from medical research to research generally and the effectiveness of the monitoring system in place to oversee whether experimentation on animals met legal and ethical requirements. It was only towards the end of the interview that the discussion moved back to the more familiar ground of 'tactics' which was here also unhelpfully personalised ('how far would *you* go'). It is not that the question of tactics is an illegitimate one, but one has to be aware of how the question of tactics is used overwhelmingly to marginalise more substantive and interesting questions that lie behind the protest in the first place.

The right to protest

Outside *Sunday Live* the question of the tactics of animal rights protesters had been, as we have seen, dominant, and continued to be so. On 15 May a *Dispatch*es documentary for Channel 4 was screened. The documentary, by David Modell, was the result of a year-long filming of activists from the Animal Liberation Front. The title of the documentary, 'Mad About Animals' accurately reflects the programme's viewpoint on the activists as it presents what appear to be an assortment of misfits, loners and the intellectually challenged. The approach is very much within the parameters of the television news frame as it focuses once more on the question of tactics (and includes one of the people who was later sentenced to jail over the removal of Gladys Hammond's body). The rather more substantive issues of the case for and against animal testing are completely ignored – but presumably that would have made for less 'dramatic' television (at least as far as commissioning editors are concerned). Press reception of the documentary agreed that it would further damage the cause of animal rights and alienate public opinion. BBC's *Question Time* (18 May) included a debate on the question of vivisection after a question from a member of the audience on whether the panel would sign the online petition that Tony Blair did. All the members of the panel agree that they would and that research for medical purposes was justified. Simon Hughes MP (Liberal Democrat) argued that it is *not* justified for other kinds of research (including for military purposes, interesting enough), while Conservative MP Ken Clarke suggested that he had looked into the question of alternatives and was convinced that they do not exist in many cases. Harriet Harman MP (New Labour) predictably gives the most pro-government and unequivocal response, accusing animal rights protesters of 'terrorizing the industry'. Interestingly, Harman uses a verb to describe the activities of the anti-vivisection campaigners. This strategy by ministers is a kind of *ideological seeding* of the public sphere which can then be subsequently nurtured by the media, in this case, converting the verb to a noun ('terrorists'). Frederick Forsyth cites the case of the thalidomide drug which caused thousands of deformities in unborn babies, but which if it had been tested on animals, would, he argues, have been stopped. Helena Kennedy, a Labour supporting member of the House of Lords cites her own professional expertise in support of animal testing and suggests that if any member of the audience had a family member who was ill, they would too. Kennedy received a big applause for her intervention, although other members of the panel received a more evenly mixed reception from the audience, judging by the visual and audio evidence. Subsequent contributions from the audience include one woman who cites the same statistic as Keith Mann, about drugs being the fourth biggest killer in the Western world and suggests that Prime Minister Tony Blair's support for the industry is all about money. Another woman suggests that more effort should be put into researching alternatives. Both thus raise issues connected to problems with 'Big Pharma' – the fact that it is a large-scale capitalist industry. The uniformity and certainty of opinion (albeit with various nuances) on the *Question Time* panel, therefore, is not universally shared by the opinion of the *Question Time* studio audience. This lack of diversity of opinion on the panel, the different status between panellists and members of the audience which the programme constructs and the lack of any questions/perspectives from the audience back to the panellists from the chair (David Dimbleby) means that while the debate certainly goes beyond the parameters constructed by the daily news bulletins, it cannot qualify as much of a contribution to public debate according to the criterion we have been using. At least, however, *Question Time* did not repeat the dominant motif of the media coverage: the issue of *tactics*. Something very interesting starts to happen to the issue of tactics, however, mid-way through the month. The *specific* tactics associated with the letter writing campaign and the removal of Gladys Hammond's body began to *mutate* into a more general questioning about the right to protest *per se* and the possible need to extend laws that can curtail and limit protest (laws which of course can be applied not only to the animal rights protesters but other campaigns as well). This slippage between specific tactics which contravene or may contravene existing laws, and the more generalised questioning of the right to protest and the need for further laws restricting the right to protest was both a media and legal construction that arguably fed into and supported each other. On 18 May, *Channel 4 News* returned to the story of Oxford University's new laboratory for animal experimentation under construction. The university, almost certainly taking advantage of the way the battle for public opinion was being lost by the animal rights activists (at least within the media), was now applying for a court order to limit protest against the building construction to four hours on a Thursday. This news item adopted a sceptical tone as to the necessity of such a draconian reduction and even the ubiquitous Iain Simpson, who is briefly interviewed, seems to agree that it is questionable. However, *Five News* (18 May) ran the same story and was far more supporting of the university's application for a court order. This support was evident in the way that it raised no questions about the further restrictions on the right to protest and the way it even linked the specific tactic of a protest on a street, using megaphones and handing out leaflets, *back* to the illegal and clearly morally unsustainable tactic of removing Gladys Hammond's body (which is referred to again, visually and verbally). If at the beginning of the month, the digging up of Gladys Hammond's body represented a morally unacceptable and extreme protest tactic, by the end of the month, the threshold of acceptable protest activity was being set an awful lot lower. Indeed, it seemed that protest using a megaphone started to be constructed, by both the media and the courts, as crossing the line of acceptable behaviour. We can track this linkage between megaphone use, the discursive construction of 'extremism' and legal judgement as it emerges and crystallises into an *image* within the public sphere. In the *Dispatches* documentary 'Mad About Animals' there is a scene where an animal rights protester during a street protest has a brief exchange with a passer-by who makes a perfectly legitimate and non-aggressive challenge to the animal rights case. The animal rights protester both continues the exchange using her megaphone, even though the passer-by is standing right in front of her, and comprehensively fails to articulate a coherent response (which amounts to 'read the leaflet'). It is to be sure, a moment which any activist in any campaign would cringe at, but this moment appears to become emblematic of 'extreme' protest (i.e. using a megaphone) within the media constructed debate. It is a testament to how important *images* are in crystallising public memory and opinion that this image of the megaphone then enters another mediated forum, Channel Five's *The Wright Stuff,* which on 18 May had a discussion about 'the rights and wrongs of protest'. The show's host, Matthew Wright, begins by reminding viewers of his own activist credentials as a student protesting against Margaret Thatcher. But he then admits to wondering about the limits of protest having watched the 'Mad About Animals' documentary. Throughout the discussion, with his panellists and members of the public phoning in and texting in, he keeps coming back to this image of using a megaphone (without specifying that it comes from the documentary) as possibly crossing the line into unacceptable protest. Sure enough, *The Daily Mail* reported towards the end of the month that the judge in the case of Oxford University's application to limit protest further outside the animal research laboratory had indeed agreed to extend the exclusion zone where protests are prohibited and *banned* what the paper called 'extremists' from using megaphones.[4] Here in the instance of the legal judgement on the megaphone and the discursive construction of the use of a megaphone as 'extremist', we can see very clearly the articulations and mutual support which legal and symbolic practice can give to each other. The final news item of the month on the issue came from Channel 4 (31 May) on the day Gladys Hammond was reburied. But it was not only Gladys Hammond who was buried in the month of May 2006, it was also any real chance of a substantive debate on the rights and wrongs of animal testing for medical research (let alone for other purposes that may or may not be going on in the laboratories).

Conclusion

This chapter has offered a chronological account of the playing out of debate and political struggle over the issue of animal rights within UK television news during the month of May 2006. We have seen that television on the whole failed to foster a meaningful contribution to public debate on the issue of vivisection. There was unequal access to the news, differential legitimation between the protagonists and a lack of dialogue between the contending positions. The narrow and overwhelming focus on the tactics of the anti-vivisection campaigners enabled the television news agenda to be captured by powerful interests in the broader media, in government, industry and wider civil society, that mobilised against the campaigners. So ingrained are the media practices of delegitimisation that the 'law and order' agenda (which treats a political issue as if it were crime news) even marginalised the presumably compelling moral case that could be made *for* vivisection. Certainly the large middle ground between the animal rights lobby and the business as usual position of industry and government remained almost wholly unexplored. That is to say that the media generally failed to exercise their capacity for independent investigative and editorial autonomy. This should be a fourth criteria added to Bennett et al.'s criteria for assessing the public sphere. This failure of editorial/investigative autonomy and the linkage between television coverage and the frames and interests of government and industry is true across the range of broadcast output for the month (its 'communicative architecture'). This suggests that ideology critique must remain central to media analysis if we are to understand how such tight parameters to public debate can be drawn across so many programmes and programme types. The agenda set by the news bulletins was generally followed by the talk and discussion programmes (where they engaged with the issue at all) and even a documentary. Indeed the *Dispatches* documentary played a key role in backing up and amplifying the dismissal of the animal rights case via a narrow focus on tactics and personalisation of the issues. No doubt, as we saw in the case of *Newsnight* and the panel on *Question Time*, the narrow social background from which many senior broadcasters and their sources are drawn helps cement what are already structural ties to the big institutions of government and industry. Over the course of the month, there was an ideological slippage within the law and order coverage, which linked animal rights protest linguistically to 'terrorism'. There was also a visual conflation whereby specific protest strategies (digging up the body of Gladys Hammond at one end of the spectrum and the use of the megaphone at the other) were blurred together in such a way that protest of any sort began to verge dangerously close to being a questionable activity. Arguably television news contributed to the context in which legal judgements were made restricting the rights of animal rights protesters to protest (judgements which may be extended to other forms of protest in the future). In only one instance, the *Sunday Live* programme, did anything remotely like a proper debate *begin* to emerge. This demonstrates that television *could* indeed play a substantive role in contributing to the formation of public opinion based on the critical, interrogative interplay of debate. This would in turn require television news agendas and approaches to be open to criticism and reframing by participants. The extent of such space for *participant reframing* should be a fifth criteria to add to the Bennett study. The *Sunday Live* programme was the only example where such participant reframing was permitted in the sample analysed. One has to conclude that the medium is a long way from fulfilling the desired norm of engaging citizens in the democratic process, whatever public bodies such as OFCOM might like to think.

8
Apathetic or Excluded? Young People, News and the Electronic Media

The aim of this chapter is to provide an empirically detailed account of how, and to what extent, young people in the UK use the media, both old and new, to access news and current affairs. This is done by summarising those examples of detailed empirical research into this subject which have been published over the past ten years, and the conclusions which the researchers have drawn from their work. Young people's use of the new media is a subject of considerable speculation and assertion, much of it ill-informed. This makes it all the more vital to try to form a clear picture of the way in which young people actually use the media to access news and current affairs.

Utopia and apocalypse

We hear a great deal today about the 'digital generation', 'cyberkids', the 'net generation' and so on. Indeed, as Vincent Mosco points out:

> Generational divisions are central to the cyberspace version of end-of-history myths. On one side of history lies a generation or two of well-meaning but old-fashioned people, at best fumbling with the new technology but not quite getting it, at worst acting like curmudgeonly sticks-in-the-mud or like Luddites fighting against the technology and clinging desperately to old, dying ways. On the other side are the children whose intellectual savvy, willingness to experiment, and youthful exuberance draw them to the new technology and the new age it represents. (Mosco 2005: 79)

Representative examples of this kind of thinking are Don Tapscott's *Growing Up Digital: The Rise of the Net Generation* (1998) and Nicholas

Negroponte's *Being Digital* (1995), although it is possible to trace the lineage of this kind of cyber-utopianism considerably further back. Thus, for example, in 1985 Frederic Golden in *Time* was proclaiming 'here come the microkids' and positing the existence of the 'computer generation' (Golden 1985: 218, 221). It is also important to note, though, that such thinking can equally give rise to dystopian visions of cyber-apocalypse. A good example of this approach is provided by Andrew Keen's *The Cult of the Amateur* in which he approvingly cites an article by an English professor, Mark Bauerlein, entitled 'Dumbest Generations: How the Digital Age Stupefies Young Americans and Jeopardizes Our Future' which apparently 'demonstrates how the Internet is making young people more and more ignorant of almost everything except online video games and the narcissism of self-authored Internet content' (2008: 212). Equally approvingly, he also quotes an article headed 'Distracted: The Erosion of Attention and the Coming Dark Age' by the *Boston Globe* columnist Maggie Jackson in which she suggests that 'the increasingly low attention span and poor cognitive skills of today's multi-tasking, digitally addicted kids threatens to return civilisation to another dark age ... of ignorance and misinformation' (ibid.: 214).

Discussion of young people's use of the new communications media, then, especially in the press and non-specialist publications, can all too easily swing wildly between hyper-pessimism and radiant optimism, with hard facts being in notably short supply. But occasionally, even in the British press, a more cautious, nuanced and better informed note is sounded. For example, in the *Telegraph*, 30 June 2007, under the atypical headline 'The myth of the MySpace generation: young Britons are far less techno-crazy than previously thought', Matt Munday noted that: 'This generation of 18- to 24-year-olds is defined by technology. They are the MySpace generation, iPod generation, digital generation, PlayStation generation – sobriquets that evoke lives spent in a techno-maelstrom of blogging, uploading, text-messaging, file-sharing, flash-mobbing and instant messaging'. However, he then went on to detail a survey of over 400 young people in 11 European countries undertaken by the global market-research company Synovate, whose Planet Edge project, which researches youth lifestyles across western Europe, paints a very different picture. According to this survey, two-thirds of 18–24-year-olds in the UK have never used MySpace; across western Europe as a whole the figure is 76 per cent. Only one in ten of British 18–24-year-olds have ever blogged, and their preferred source of new music turns out not to be MySpace, iTunes, MTV, nor even ripping their friends' music collections to their own hard drives, but the radio, which is cited by 31 per cent of British teenagers, with the Internet, friends and MTV each cited by 20 per cent of respondents.

Planet Edge fitted their respondents into one of three categories: 27 per cent were 'cybernauts' – young people who like to be ahead of the game in terms of technology. However, the majority, 53 per cent, do not love technology per se, but view it simply as a facilitator which helps them to communicate or to entertain themselves. They thus tend to use it in a largely functional way, such as emailing, banking or shopping online. Finally there are the 20 per cent described as 'digital dissidents' who actively dislike using technology and avoid it wherever possible. However, the proportion of 18–24-year-olds who are less than techno-savvy is likely to be even greater than the survey's figures suggest, since all the research was done in an online panel, and it is thus fair to assume that the respondents came from the ranks of technologically adept.

The figures are fairly consistent across western Europe, and do not show huge disparities between young people's experiences of the new communications technologies in urban, suburban or rural areas. The picture that emerges, then, is not of a generation unified by technological change but actually divided by it – with less than a third in the fast lane of the digital superhighway. It also appears that class is a significant factor in digital take-up and use. Thus Julian Rolfe, the manager of Planet Edge, is quoted to the effect that

> We found that those from the middle class and above understand the value of the internet, and tend to have it explained to them by parents, by older generations and in school. As soon as you get underneath that, the knowledge is much less likely to be passed down at home, nor are they as likely to use computers in the workplace. And for many, IT at school is pretty basic. (Munday 2007)

The article also quotes David Buckingham, who cautions that

> On one level, I want to say that all this talk of a MySpace generation is not as dramatic as everyone is making out. But we should be careful of saying 'We've seen it all before' or 'Nothing has really changed'. Things are evolving very quickly. The issue is how lasting any of these innovations will be. We probably won't have MySpace in ten years' time, although we will have something similar but different. Labelling a generation is part of a broader thing of seeing young people as somehow exotic and different. They're labels that reflect this mixture of anxiety, but also hope, about a younger generation. (Ibid.)

And, crucially, these are not labels which are actually used by young people to describe themselves.

Anyone who sets out to discover the hard facts about what media young people use and, in particular, whether they use them to access news and current affairs will soon discover that recent and reliable empirical evidence from the UK is in fact pretty thin on the ground. In this chapter we have drawn on work undertaken by David Buckingham in 1996 and 1997, which is detailed in Buckingham (2000a, 2000b); *New News, Old News*, a study undertaken in 2002 by the Broadcasting Standards Commission and Independent Television Commission; *UK Children Go Online* (*UKCGO*), a project conducted by Sonia Livingstone and Magdalena Bober at the London School of Economics between 2003 and 2005; *New News, Future News*, a study undertaken by OFCOM in 2006; OFCOM's 2008 *Media Literacy Audit*; and an ongoing study of *Newsround* and its audience being conducted by researchers at the universities of Cardiff, Bournemouth and Coleraine.

News sources

If we begin by looking simply at the time devoted to the Internet and television by young people, we discover that the latter is clearly the winner. Hours devoted to the Internet are clearly increasing, but not, apparently, at the expense of those spent watching television (see Table 8.1).

Moving on to the question of how young people access news, we again find that television comes out on top (see Table 8.2).

However, it should be noted that the young people in Table 8.2 are of a different age group from those in Table 8.1. Furthermore, as OFCOM itself admits (2007b: 70–1), the research base used in 2006 was different from that used in 2002, which makes for certain methodological

Table 8.1 Weekly media consumption, 2005, 2007 in hours

	Television		Internet	
	Hours	No. of young people	Hours	No. of young people
8–11, 2005	13.2	768	3.8	315
8–11, 2007	15.4	1336	7.8	875
12–15, 2005	14.7	761	7.1	416
12–15, 2007	17.6	1352	13.8	1045

Source: OFCOM 2008a: 23.

Table 8.2 Sources used for news, in percentages

	Television	Newspapers	Radio	Internet	Magazines	Mobile
All 2002 (4,662 respondents)	91	73	59	15	13	
All 2006 (2,216)	94	67	52	27	17	3
16–24s 2002 (217)	92	78	60	26	18	N/A
16–24s 2006 (352)	86	61	44	35	23	N/A

Source: OFCOM 2007b: 25.

problems, and this needs to be borne in mind when reading the relevant comparative data below. But, bearing the latter in mind, it does appear as if young people's use of the various media in order to access news is not significantly different from that of the population as a whole, although it should be noted that less young people used television and more used the Internet in 2006 compared with 2002. The OFCOM research also reveals that

> The average number of news platforms used was 3.8, although 30 per cent of the population used over five platforms for news. The number of platforms used for news varied by age and socio-economic group, which is likely to be because of the availability of such platforms within the groups – the over-65s and the 16–24s used a lower than average number of platforms for news (3.3 and 3.5 respectively), as did DEs (3.2) compared to ABs (4.6). (Ibid.: 27)

Research carried out for the *Newsround* study between January and April 2008 at 8 primary and secondary schools in Glasgow, Cardiff, Bournemouth and Coleraine, discovered that among the 187 young people who named their favourite sources of news, the results were as follows:

		Per cent
Television:	112	(51.9%)
Newspapers:	15	(6.9%)
Internet:	15	(6.9%)
Radio:	14	(6.5%)
Friends:	12	(5.6%)
Friends and parents:	8	(3.7%)
TV and Internet:	8	(3.7%)

Table 8.3 Televisual vs non-televisual news sources

	Primary school (104 young people, 8–12)	Secondary school (112 young people, 13–15)	Total answering
TV	76 52.1%	70 47.9%	146
Non-TV	28 40%	42 60%	70

Source: Carter et al. 2009: 15.

Table 8.4 The main role of television should be to inform and educate

	Strongly agree	Slightly agree	Don't know	Disagree
Boys 12–15 (512 respondents)	26	46	18	10
Girls 12–15 (505)	23	48	14	15

Source: OFCOM 2008b: 32.

Television thus wins out not simply over individual non-televisual sources but over all of the latter combined (Carter et al. 2009: 15). Furthermore, the research showed that in terms of their preference for television as a news source there was little difference between young people at primary and secondary school (see Table 8.3).

How much television news?

OFCOM's 2008 Media Literacy Audit asked young people whether they agreed or disagreed with the statement that the main role of television should be to inform and educate. The results were as shown in Table 8.4.

If we aggregate those who strongly agree and slightly agree this would suggest that television news should appeal to a significant proportion of young people. However, the available research tends to suggest otherwise.

Although a great deal of comment in the contemporary media bemoans the lack of time which young people apparently spend watching or listening to the news, this is hardly a new complaint. As always, the golden age is hard to locate. Thus Asa Briggs noted BBC audience research in the 1950s which found that 16–19-year-olds 'tended to see and hear less news than their elders. And when they saw or heard, they were more "desultory", being prepared to switch attention, if not switch off, after the headlines' (1995: 72). *New News, Old News* argues that 'audiences for news have always been overwhelmingly concentrated among the older age groups' (Hargreaves and Thomas 2002: 30). And David Buckingham notes that amongst the young people whom he interviewed in Philadelphia and London in 1996 and 1997

> Despite the differences between the various groups of students whom I interviewed, few expressed great enthusiasm for television news. The programmes nominated as favourites in our warm-up discussions were almost exclusively comedies, soaps, movies and sport... With a few exceptions, expressions of even moderate interest in news were only to be found among the oldest age group. More frequently, news was rejected as simply 'boring', and in several cases exaggerated groans greeted my introduction of the topic... In general, news was perceived as repetitive and lacking in entertainment value. (Buckingham 2000a: 64)

But as Buckingham also notes: 'There has been very little research on young people's relationship with television news... evidence on this topic is exceptionally sparse' (ibid.: 9). To a significant extent, this remains the case.

According to *New News, Future News*, in 2001 16–24-year-olds watched approximately 45 hours of news per year, compared with more than 180 hours amongst those over 55. And of those who never watched current affairs, almost one in three (30 per cent) were under 35, making this the biggest age group of those who simply ignore such programmes. However, the study also reported that 56 per cent of young people picked up topical information and news from unconventional sources such as comedy shows, and 29 per cent from chat shows. The later study, *New News, Future News*, reported that in 2006:

> Actual consumption of news by the 16–24 age group has decreased across all themain news platforms, with the exception of the internet and magazines, which have increased. The average amount of television news watched by 16 to 24 year olds is around 40 hours a year – or around 45 minutes a week – compared to around 90 hours a year for the wider population. (OFCOM 2007a: 61)

In addition, 50 per cent of young people reported that they followed the news only when something important or interesting was happening; this compares with 33 per cent in 2002. Comparing these figures with those for the total sample, we find that 32 per cent of all respondents in 2006 said that they followed the news only when something important or interesting was happening, and 26 per cent in 2002.

New News, Future News also draws on data from BARB to paint a revealing picture of broadcast news viewing during 2006. Thus, during the week commencing 10 October, 45 per cent of all individuals watched more than one news programme on the terrestrial channels, and 15 per cent watched only one. Amongst the 16–24-year-olds, 17 per cent watched more than one news programme, and 19 per cent only one. BARB data are also used to reveal that in 2001, 5 per cent of the news programme audience was composed of children and 6 per cent of 16–24-year-olds; by 2006 both figures had dropped to 4 per cent. Interestingly, the news programme which in 2006 had the highest percentages of viewers who were children (6 per cent) and 16–24-year-olds (9 per cent) was *Channel 4 News*. Four per cent of the audience for *Sky News* was made up of children (which is the same figure for the BBC early evening news and ITV late evening news), and 4 per cent was made up of 16–24-year-olds. In the case of *BBC News 24*, 3 per cent of its audience was made up of children (the same figure as the BBC late evening news) and 4 per cent of 16–24-year-olds.

Attitudes to television news

In certain important respects, younger and older people differ when it comes to the news topics in which they are most interested. As Table 8.5 shows, younger people tend to be less interested in current events and politics at all levels than older people, and more interested in entertainment, sports and celebrity behaviour (although not human interest stories, travel and consumer affairs). The research also found that news about current affairs was frequently consumed only passively – for example, a television or radio switched on by their parents, news in music, sport or other radio programmes, or headlines which they happened to see.

When we turn to televisual news specifically, one of the main sources of complaint amongst young people concerns what they perceive as its lack of relevance to them. According to *New News, Future News*, 44 per cent of young people felt that this was true of television news in 2002, and 64 per cent in 2006. This compares with 34 per cent of

Table 8.5 Interest in news topics by age (in percentages)

	Total (2,216)	16–24 (352)	25–44 (917)	45–64 (575)	65+ (372)
Current events in the UK	55	41	56	58	59
Current events in my region	50	30	51	53	58
Current local events where I live	48	31	45	48	58
Weather	48	37	51	50	50
Crime	47	52	51	49	34
World-wide politics and current events	41	26	40	47	46
Sports	39	43	38	38	38
Human interest stories	38	27	38	42	38
UK-wide politics	37	20	37	43	38
Entertainment	34	48	38	28	27
Politics in my region	28	17	23	24	34
Travel	26	15	25	30	30
Consumer affairs	23	8	21	30	28
City, business and financial issues	19	10	18	25	18
Celebrity behaviour	13	20	16	8	9

Source: OFCOM 2007b: 16.

all people surveyed in 2002, and 55 per cent in 2006. Clearly there is a significant generational gap here, but equally clearly there appears to be a significant degree of alienation from this aspect of television news within the population as a whole.

The qualitative research undertaken for the *New News, Future News* study found that those young people who judged television news as irrelevant did so on the grounds of both content and form. In terms of the former, many argued that they were not interested in news which, they felt, had no impact on them directly, and this tended to include politics and current affairs. Thus a white 16–18-year-old from Bradford stated that: 'Politics; you just don't want to know. Because it does not affect us', whilst a white 18–24-year-old from Cardiff maintained: 'I am only interested in the news that affects me, for example the proposed smoking ban' (OFCOM 2007b: 54). In terms of form, the youngest respondents found it hard to relate to the style and tone of much news reporting which, they felt, not only excluded them, but also made the news very much the domain of the older generations. Indeed, the young people interviewed for this project clearly recognised that their lack of interest in what they perceived to be 'serious news' was something which defined them as being different from older people: 'Younger people saw not being interested in this type of news as a characteristic of their stage of life and general outlook – almost as a positive affirmation that they were behaving as people of their age should' (ibid.: 21). Thus a white 16–18-year-old from Bradford argued that: 'There is a lot more interest from older people than people our age because we don't have time. We've got to be out' (ibid.: 52). Respondents tended to imply that when they were older they would have more interest in 'serious' news. Thus an 18–24-year-old Indian Hindu/Sikh from Leicester opined: 'As you grow older, you start watching more news' (ibid.: 53). Many of the young people interviewed felt that as they had no adult responsibilities, they did not need to know what was happening in the world. There was also a sense of peer pressure at work here, since several talked about there being a social stigma attached to admitting interest in the news. Thus a white 18–24-year-old from Cardiff stated: 'It is not cool to watch the news. There's better things going on' (ibid.: 52).

This sense of apathy and disconnection did not, however, apply to all aspects of the news – as noted above, there was interest in news about sports and celebrities, and in other news which was needed to engage with peer groups and youth culture. There was also interest in both major national and international events such as 7/7 and 9/11, and in local news concerning people or places which they knew. However, a number of programmes were mentioned as offering news in styles and formats which appealed to this age group, and these included *Newsround*, which was liked for its straightforward reporting of stories and *The Big Breakfast* which was felt to be aimed at younger audiences. As a white 19–24-year-old from Glasgow put it: 'It was for young people and that's probably why we all liked it. It was funny and it lightened up the news a wee bit. It made it a wee bit more amusing whereas it seems to be quite depressing' (ibid.: 54).

As well as a sense of apathy and disconnection amongst young people, an attitude of cynicism towards the way in which the news was reported, by all the British media, was also discovered amongst some of the young people whose views were gathered for *New News, Future News*. In fact, this was evident amongst all sections of the audience but was most apparent amongst those under 25. According to the report: 'Whilst the respondents acknowledged that the facts of the stories were likely to be accurate, they expressed reservations about the editorial policies and controls that lay behind story selection and presentation. Most news platforms were perceived to be affiliated to or controlled by someone with vested interests' and were thus easily seen by young people as an extension of authority which was trying to control them.

Table 8.6 Trust in news programmes

	Always	Mostly	Neither believe nor disbelieve	Not much	Not at all	Don't know
8–11	38	41	10	4	1	6
12–15	34	49	9	4	2	2

Source: OFCOM 2008b: 35.

Thus a white 18–24-year-old from Cardiff complained that 'the BBC is basically run by the government so you'll only hear on there what the government want you to hear' (ibid.: 53). This cynicism tended to fuel the apathy and disconnection noted above in that if the news was regarded as inherently untrustworthy, there was little point in engaging with it. However, there was significantly less cynicism about local news coverage, because it was felt to state facts and, more importantly, because respondents felt they had the ability to verify the version of events presented. As a white 18–25-year-old from Cardiff put it: 'They (local news) are not trying to influence people. It is just fact. Something happened in this village ... in that village' (ibid.).

On the other hand, it should be pointed out that when the OFCOM 2008 *Media Literacy Audit* posed the question 'Do you believe what you see on news programmes?' it discovered relatively high levels of trust amongst young people aged 8–15, although the figures may possibly suggest that trust declines with age (see Table 8.6).

Young people, television news and politics

From the above, it is clear that many young people are uninterested in, when not positively hostile to, news about politics on television. From here, of course, it is but a short step to either (a) blaming the media for presenting politics in such a way as to alienate young people from the political process; or (b) blaming the young for being too stupid or apathetic to take an interest in news about politics on television. A further extension of (b) is to blame the 'tabloidisation' or 'dumbing down' of television on young people's alleged inability or lack of desire to watch anything other than superficial and sensational entertainment programmes. However, another approach is possible, one which understands young people not as apathetic but as *antipathetic* – not to politics per se but to party politics on the Westminster model and, in addition, to the manner in which this shapes and dominates the way in which political issues and events are represented on television. This, admittedly, is a version of (a), but, crucially, it is one which accepts that the political process is *itself* alienating (and by no means simply to young people), and puts the blame for young people's disaffection on both the broadcasters and the politicians. Thus, as David Buckingham suggests

> Young people's apparent rejection of politics and of news media could also be seen to reflect their sense of *exclusion* from the domain of politics, and from dominant forms of political discourse. In this situation, politics and government remain at best an abstraction, and young people's lack of interest in politics is merely a rational response to their own powerlessness. (Buckingham 2000b: 172)

From this perspective, young people can be seen as politically disenfranchised, with the biggest failure being on the part of the broadcasters and the politicians to connect with the kind of 'everyday politics' or small 'p' politics which are of most interest to young people (and, for that matter, to many older people too). In this process, schools, too, must shoulder a fair share of blame for not teaching young people adequately about either Westminster politics or politics in the wider sense – the result of decades of conservatively minded terror that young people, and especially working class ones, might be 'indoctrinated' by 'left wing' teachers. This has left a gaping void in their education and socialisation which no amount of citizenship classes can possibly fill.

Both in the UK and the US, Buckingham found that the young people whom he interviewed, aged 11–17, were, on one level, extremely cynical about politics as conventionally defined – that is, about the actions of politicians, who were often condemned as not merely boring but also corrupt, uncaring, insincere and self-interested. They were also criticised for their neglect of issues specifically related to young people and for what were seen as authoritarian policies in areas such as education and youth crime. Party politics was widely dismissed as a kind of dishonest game which had little relevance to their everyday lives and concerns. However, this does not mean that they are not interested in politics in a wider sense. As Buckingham argues:

> Children develop 'political' concepts at a very early stage, through their everyday experiences of institutions such as the school and the family: notions of authority, fairness and justice, power and control, are all formed long before they are required to express their views in the form of voting. The choice available at school lunches, the

> attempt to introduce compulsory uniforms, or even the organisation of the school playground – which were among the topics discussed with considerable passion by the children in my interviews – are, in this respect, just as political as what goes on in parliament. One might well make a similar case about sports and entertainment: the success of Tiger Woods or the Spice Girls can clearly be interpreted as 'political' phenomena, as they implicitly were by some of the children here. (Ibid.: 177)

From this perspective, young people experience considerable difficulty in connecting the 'political' dimensions of their own everyday experiences with the official discourses of politics, which in the UK means Westminster politics, as encountered in broadcast news programmes. In particular, there is a huge gap between the ways in which political issues relating to young people are framed within the news and the ways in which these issues are experienced and perceived by young people themselves. As Buckingham concludes, the challenge for journalists as well as for all those working with young people is to

> find ways of establishing the *relevance* of politics, and of *connecting* the 'micro-politics' of personal experience with the 'macro-politics' of the public sphere. This will not be accomplished simply by dumping information on young people, or indeed by issuing them with implicit injunctions to do their civic duty; and it will require a definition of politics that goes well beyond the formal operations of political institutions. (Ibid.: 187)

Indeed, this applies equally to journalism aimed at adults as at young people, as does Buckingham's judgement that

> Much greater efforts need to be made not merely to explain the causes and the context of news events, but also to enable viewers to perceive their relevance to their own everyday lives. News can no longer afford to confine itself to the words and actions of the powerful, or to the narrow and exclusive discourses which currently dominate the public sphere of social and political debate. (Ibid.: 182)

In his research in Britain, Buckingham asked young people about their attitudes to three news programmes made specifically for their age group: *Newsround* (BBC), *First Edition* (C4), and *Wise Up* (C4). (Significantly, the only one of these still running is *Newsround,* which began in 1972. *First Edition* ran from 1994 to 2002, and *Wise Up* from 1995 to 2000.) What he discovered was that young people do not like to be talked down to or to be patronised by programmes trying to be 'cool' or 'hip'. Young viewers liked to be presented with information that was relevant to their own everyday concerns, or which concerned matters about which they did not already know, as opposed to merely a simplified or more 'entertaining' version of the mainstream news. They also liked items which featured young people presenting their own points of view, or simply 'ordinary' people having their say.

Highly critical of the all-too-common assumption that the only thing that young people want from television is to be entertained, and insistent that they also want to be informed and helped to think about the world around them, Buckingham also takes pains to stress that: 'For all its shortcomings, news journalism remains the primary means of access to the public sphere of political debate and activity' (ibid.: 186), and that it does so for the young, middle aged and elderly alike. He concludes that his research,

> clearly confirms the need for innovation if news is to reawaken the interest of younger audiences, and indeed of the large majority of the population. News is undeniably one of the most conservative of media genres, and the horrified charges of 'dumbing down' that typically greet even the most marginal changes in its approach attest to the enormous symbolic importance that seems to be invested in its curiously limited forms and rituals. (Ibid.: 182)

The young people who took part in Buckingham's research clearly felt very strongly that young people should be heard and represented in the media on their own terms. Buckingham himself obviously sympathises with this view, and in particular stresses that if young people are to perceive news as relevant to them, there is a need to rethink not only traditional notions of what counts as news but also the formal strategies of news production and presentation.

At the same time, however, Buckingham is rightly critical of the approach taken by cultural studies academics such as John Fiske to the problem of young people's disengagement from television news and current affairs programmes, an approach which concentrates almost entirely on the micro-politics of cultural consumption (television viewing very much included), the apparent openness of cultural texts to 'popular' and even 'subversive' readings, and the pleasures, indeed 'empowerment', to be gained from such readings. For Fiske, the answer to the problem of disengagement is for television news to become more popular by becoming more pleasurable, taking on more of the characteristics associated with tabloid journalism, with fictional genres such as soap opera, and with other popular forms of entertainment. However, it is difficult not to regard Fiske's approach as coming perilously close to regarding news as important only insofar as it relates to 'everyday politics' and ignoring the other functions which it should perform in a democratic society consisting of well-informed citizens. As Buckingham puts it, 'it is hard to see how everyday lived experience can be conceptualised in "political" terms without the ability to connect it to the wider world of collective action – and hence without access to *information* about that wider world' (ibid.: 186). Furthermore, in relation to young people in particular, television news has, or at least should have, particular educational responsibilities, which Fiske's subjectively idealist approach, institutionally and socially deracinated as it is, simply ignores. (For a trenchant critique of this kind of thinking see McGuigan 1992: 70–5).

Young people, the Internet and politics

When Buckingham was writing about his research into young people's news viewing, the Internet was not as developed or as widely used as it is now; nonetheless, hopes were already being commonly expressed that it might lead to greater political awareness, encourage wider participation in the democratic process, and open up a new public sphere in which previously silenced or marginalised voices could be heard. However, Buckingham presciently warned that not only do such arguments tend to underestimate the continuing relevance of more traditional forms of communication but that they also tend to encourage the delusion that new forms of communication will *of themselves* create new opportunities for forms of democratic communication. As he put it: 'The new forms of cultural expression envisaged by enthusiasts for digital media will not simply arise of their own accord, or as a guaranteed consequence of technological change; we will need to devise imaginative forms of cultural policy that will foster and support them' (2000b: 222). Thus far, these policies have signally failed to appear, and information on the extent to which young people use the Internet to access news and other forms of information about current affairs remains thin on the ground. There exists a substantial and ever-growing literature and para-literature on the dangers posed to young people by the Internet, but much less of any substance on the democratic opportunities which it offers, or could offer.

When asked if they agreed that the main role of the Internet should be to inform and educate people, 44 per cent of 12–15-year-olds strongly agreed, 39 per cent slightly agreed, 12 per cent did not know, and 5 per cent disagreed (OFCOM 2008b: 64). Thus, as in the case of television (see Table 8.4), one might expect the Internet to be a popular and attractive source of news and information about current affairs for young people. However, the available evidence suggests that although the Internet plays an increasingly important role in young people's social and school lives, it has yet to play a similar role in their civic and political lives. The explanation for this may lie equally in their attitudes to civic and political engagement, and in the nature of the opportunities for such engagement offered to them by the Internet.

The *New News, Future News* project reported that according to the Nielsen//NetRatings, in September 2006, Internet users spent 40 minutes per month using news and information websites, and 29 minutes per month using websites featuring current news and global events. For 2–15-year-olds the figures were 12 and 5, and for 16–24-year-olds 23 and 10. (However, it should be noted that these figures exclude access at libraries and universities.) In the same month visits to specific news sites by young people were as shown in Table 8.7.

According to the qualitative research carried out for this project, the use of the Internet for news appeared to be almost incidental for younger people. Thus a white 16–18-year-old from Leeds stated that: 'When I log onto the internet, the first page that comes up has a little news bit on it. Sometimes a little picture pops up with a little headline, and if it's interesting, I'll click on it and have a little look', and a white 19–24-year-old from Glasgow said that: 'Usually, on Yahoo, when I'm searching for something, the news headlines are down the side, so if there's something interesting then I'll have a look' (ibid.: 32). In the light of some of the young people's remarks about television news being

Table 8.7 Percentages of young visitors to major news websites

	BBC News	CNN	Yahoo! News	Google News	Guardian Unlimited	Telegraph
2–15	8	8	4	8	5	2
16–24	10	16	10	13	11	6

Source: OFCOM 2007b: 92.

the voice of authority, it is interesting that the report reveals that

> some felt that the Internet offered a less biased alternative. Those with access to the internet felt, when getting news from this source, they were 'in control' and 'nobody could control their minds'. These views were based on the perception that they could access the stories they wanted (as opposed to being told the ones others deemed important) and consume as much or as little as they pleased.

In this respect the report quotes an African 18–24-year-old to the effect that 'I am quite sceptical about my sources…it can be modified to what the public gets. I go to the internet to get different sources and then digest' (ibid.: 53).

The most comprehensive analysis of the extent to which young people in the UK use the Internet to access news and current affairs online is to be found in the *UK Children Go Online* project. This investigated 9–19-year-olds' use of the Internet, and combined qualitative interviews and observations with a major national face-to-face survey of some 1511 young people (users and non-users) and their parents. The most useful accounts of the news and current affairs aspect of the project are to be found in Livingstone (2007), Livingstone, Couldry and Markham (2007) and Livingstone (2009).

Livingstone reports that 'over and over again, the conversation flagged when we switched from communicating with friends to the idea of communicating online in order to connect to the world of politics' (2007: 108), thereby neatly replicating Buckingham's experiences when he introduced the topic of television news into his discussions with young people. And although it is clear that the vast majority of young people who use the Internet seek out information of one kind or another, only a minority do so in order to access news (see Table 8.8).

In attempting to understand the extent to which young people use the Internet for purposes other than entertainment, Livingstone investigated their use, and non-use, of civic websites – the word 'civic' being chosen deliberately in order to avoid the negative connotations for many of the word 'political' (see Table 8.9).

Here some interesting demographic differences became apparent. For example, 31 per cent of girls visited charity sites, compared with 22 per cent of boys; 35 per cent of older teens (16–19-year-olds) visited such sites, compared with 20 per cent of 12–15-year-olds. In the case of human/gay/children's rights sites, 23 per cent of older teens visited these, compared with 13 per cent of 12–15-year-olds. In more general

Table 8.8 Demographics of Internet use for seeking information, in percentages

	Seek information (not for school)	Seek advice online	Look for news online
All	94	25	26
Boys	94	26	28
Girls	94	23	22
ABC1	95	26	28
C2DE	93	23	22
9–11	89	N/A	N/A
12–15	94	21	17
16–17	96	29	34
18–19	97	32	41

Source: Livingstone 2007: 114.

terms, 54 per cent of 12- to 19-year-olds who went online at least once a week visited at least one such site. Girls and young middle-class teenagers tend to visit a broader range of civic sites, and the breadth of civic sites visited also increases steadily with age. Thus factors such as gender, age and social class (not to mention social capital) all play a key role in whether young people use the Internet to visit and interact with civic websites:

> Young people's motivation to pursue civic interests online depends on their background and their socialisation, and it is not greatly affected by the amounts of time spent or levels of expertise online. Rather, those with prior civic or political interests find the Internet a useful resource for pursuing those interests; similarly, those motivated to explore the Internet creatively do so, resulting in an active and creative engagement with the medium, but not necessarily drawing them into greater civic or political engagement than before. In short, interaction and civic engagement are not to be regarded as sequenced 'steps' on a 'ladder' of participation from minimal to more ambitious modes of participation. (Livingstone, Couldry and Markham 2007: 24)

Clearly, then, visiting civic websites is low on young people's priorities, and when they do actually visit they tend to use them more as a source of information than an opportunity to become engaged politically, in the broad sense of the term. However, Livingstone, like

Table 8.9 Visiting or not visiting civic websites, in percentages

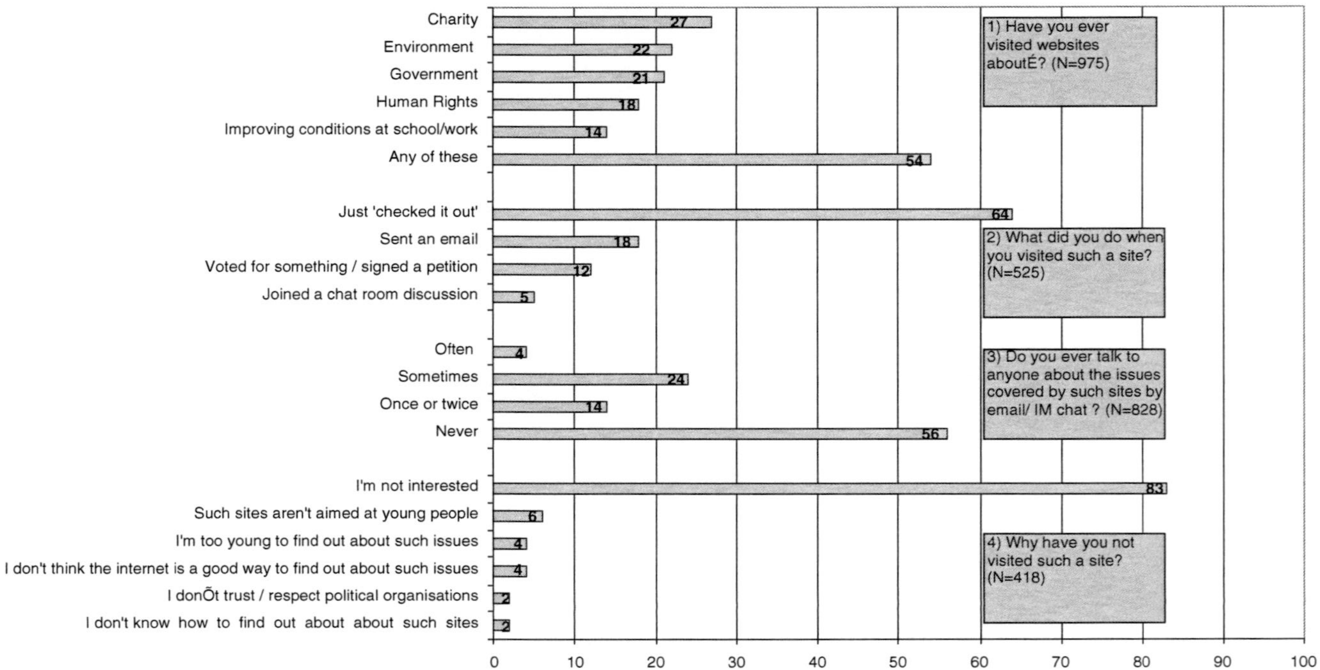

Source: Livingstone 2007: 116.

Buckingham in the case of young people and television news, suggests that this should not be taken as evidence that young people are simply uninterested in politics and political communication per se but, rather, as an indication that they are not attracted or engaged by the ways in which these are presented by the Internet. She also notes that although politicians and Internet content producers alike repeatedly stress the importance of 'having your say', young people realise all too well that this is not the same thing as 'being listened to', both online and offline. Young people are not exactly encouraged, except rhetorically, to participate in the shaping of the society in which they live, and this form of democratic deficit is mirrored by what Livingstone describes as a communicative deficit in both the traditional and new media. She thus concludes that

> Young people consider the online invitation to be false – no adult is listening or responding, and decisions are taken elsewhere. Perhaps, in consequence, it also seems that while these sites offer young people their 'right' to be heard, they generate little if any sense of any accompanying responsibility to participate... One is tempted to suggest that it is those making the invitation, not those responding to it, that lack the motivation to participate. (Ibid.: 120–1)

Conclusion

Although not all of the findings of the different pieces of research presented above are necessarily consistent with one another (on levels of trust in television news, for example), there are nonetheless broad areas of agreement. In particular it is clear that younger people watch less television news than do older people, but, there again, they watch less television *tout court*. It is also clear that disengagement from television news and current affairs programmes about politics cannot be separated from disengagement from the political process as a whole, but (a) this does not mean that young people are not interested in small 'p' political issues; and (b) this is a characteristic which they share with a seemingly ever-increasing number of older people. The solution to this problem would seem to lie more in trying to extract television news and current affairs programmes from the airless and incestuous 'Westminster bubble' than in exhorting young people to buck up their civic ideas. In terms of the Internet, its much vaunted (and indeed promising) democratic potential as a new public sphere has yet to be realised as far as most young people are concerned, and merely ensuring that everyone can access the Internet via broadband (which appears to be current government policy) will do nothing to affect the ways in which young people, or indeed people of any age, actually use it. As almost all of the studies analysed in this chapter indicate, factors such as gender and social class, as well as age, play a very considerable role when it comes to the media which people use, and, in particular, the purposes for which they use those media. Narrowing the digital divide, then, is not simply a matter of laying reams of cable and facilitating Freeview, but also of addressing the gaping social and cultural divides which so disfigure the contemporary UK. Merely providing people, of any age, with the latest communications technology will do absolutely nothing at all to help them to use it to become better informed and more active citizens.

9
Talk Back: Young Audiences and Reception

Introduction

Are younger audiences interested in politics? Do they feel represented? Are media images and messages concerning young people relevant to their everyday lived experience? If not, what can be changed? There are a number of intersecting debates circulating in academic, political and broadcast arenas that have underpinned the research study discussed in this book. Concerns centre on young people's apparent declining interest in the political process or at least traditional notions of 'Westminster-based' politics and fears that they are more likely to vote for reality TV contestants than in a general election (Lewis, Inthorn and Wahl-Jorgensen 2005). Yet as we discussed in Chapter 4, there is evidence that young audiences *are* interested in aspects of a mainstream news agenda (although not Westminster-based politics); that they also have a distinct set of interests that are particular to being young; that they are interested in 'single issue' politics (lobbying for change around interest-linked campaigns such as world poverty, the environment, animal rights), and that, yes, they are also inclined towards what might be called celebrity or entertainment news. We have also highlighted that young people tend to be criminalised within mainstream television news (see Chapter 5); yet ironically it is precisely these same marginalised audiences that represent the future survival of television news and within the broadcast industry there is considerable discussion about how best to attract and maintain these viewers.

In such debates younger audiences are envisaged in specific ways and much is assumed about their relationship with media and politics not least in terms of 'what audiences want'. Yet it is surely important that we explore actual audiences and how they engage with current provision of TV news and politics. Is there evidence of apathy or alienation from the political process? To what extent are their views shaped by social class or ethnicity? This chapter explores younger audiences and their engagement with existing television news and current affairs programming. There is often a significant 'gap' between television production personnel and their audiences and this phase of the study was designed to address this with a view to shedding light on how this vital target group might themselves approach the problem of disconnection.

Sample and methods

To explore these issues we held a series of focus groups with young people living in London and the South East of England (Sussex and Essex). A total of 12 focus group sessions took place in youth/community centres, university premises, participants' homes and we also convened a session in a remand centre for young offenders. As illustrated in Table 9.1, we included a total of 80 participants most of whom were aged 20 or under (n = 45). The gender mix was reasonably even (male = 46; female = 34) and the sampling was purposive to cover a range of demographic characteristics such as age, ethnicity and social class. Thus, sessions involved young people from low-income backgrounds who attended a youth club in North West London and more affluent sixth form students from a highly selective grammar school in Essex. We did not specifically target students in higher education but in fact all of those in our mid-high income groups were attending university or sixth form college as were some of those in lower-income groups. Participants came from diverse backgrounds as follows: 'white' (n = 47); 'black' (n = 15); 'Asian' (n = 13); 'mixed race' (n = 4) and Hispanic (n = 1).

We decided to recruit 'naturally occurring' groups: that is, people who would normally meet or socialise together and therefore knew each other prior to the session. Media is discussed in everyday life and we were keen to maximise social interaction within the sessions. Thus our focus groups were designed to address participants' views concerning television news and politics but also to tap into the jokes, banter and anecdotes that typify discussions between groups of friends or school mates but are traditionally very difficult to reproduce within a research session (Carter and Henderson 2005). A related aim was that we were particularly keen to involve those from 'hard-to-reach' groups and holding a group session in a convenient location with friends/peers helped facilitate a more relaxed session. The sessions were tape recorded and fully transcribed. The sessions were conducted during May to August 2007. This period coincides with the resignation of Labour Prime Minister Tony Blair and former Chancellor Gordon Brown taking over his role on 27 June 2007.

Group protocol

The sessions had three main phases. First, participants completed an individual questionnaire designed to capture demographic details; their media consumption, political interest and voting intentions. Second, participants were split into smaller groups, given a set of still photographs taken from television reports (that featured in our news sample) and asked to reproduce a story they might see on the news. Third, the group then came together to present and discuss their news reports and more general issues concerning representations of young people, politics and their preferences for media coverage. This format has been used in the past to explore public understandings of industrial disputes, international conflict and health/social issues (Glasgow Media Group 1976, 1980; Henderson 2007; Philo 1990).

Media consumption, sources of political information and voting intentions

We were keen to know which media sources our participants regularly consulted for news and asked 'Where do you go first for news?' Interestingly, despite the common assumption that younger people are mainly using new media, we found that 'old' media still had a crucial role to play (as discussed in the previous chapter). Indeed well over half of our sample went first to 'TV' (53/80). In fact far fewer than expected went first to the Internet (17/80) and of those who did most came from more affluent backgrounds and tended to be in higher education. The responses to using the Internet were fairly polarised with 20/80 people using it 'most days' and 35/80 using it 'rarely' or indeed even 'never' for news. In addition, we found that newspapers also featured fairly highly as a first source (13/80). In terms of television news, we also asked about their frequency of consumption and found that more than half of our sample said they watched television news 'every day' or 'at least 3 days' per week. When asked about news channels over half of our participants (n = 41) watched 'BBC news' describing it as 'trusted' and 'reliable'. Other popular channels were *ITV, Sky* and *Channel 4* (partly perhaps due to scheduling as university students, Group 6 described 'You know

Table 9.1 Description of the focus groups

	Description		No. of participants			
Group ID	**Type of group**	**Location**	**Female**	**Male**	**Age range**	**Ethnicity**
FG1	State school students Low income	Hillingdon, Greater London	7	4	16–18	8 White 1 Black 1 Asian 1 Mixed race
FG2	Youth group Low income	Hammersmith, West London	3	5	18–24	3 White 3 Black 1 Asian 1 Hispanic
FG3	State school students Low income	Uxbridge, Greater London	3	3	16–17	3 White 1 Black 2 Asian
FG4	Community group Low income	Kilburn, NW London	1	7	18–22	6 Black 1 Asian 1 Mixed race
FG5	Community group Low income	Brent, NW London	3	3	16–24	2 Black 4 Asian
FG6	University students Middle income	Bounds Green, North London	4	3	21–24	7 White
FG7	Friendship group 1 Middle income	Brighton	4	3	19–21	6 White 1 Asian

FG8	Friendship group 2 Middle income	Brighton	4	2	17–24	6 White 1 Black
FG9	Group of young offenders Low income	Greater London	0	5	22–24	2 White 1 Black 2 Asian
FG10	Grammar school students Mid-high income	Essex	0	5	16–17	5 White
FG11	Friendship group/University students Mid-high income	Colchester, Essex	0	6	16–20	5 White 1 Mixed race
FG12	Grammar school students Mid-high income	Essex	4	0	16–17	4 White

why we all watch Channel 4 News though, don't you? Because it's on after *Hollyoaks* (laughs). That is why I watch it!' and another added 'I watch (lunchtime) BBC News waiting for *Neighbours*)'.

But other diverse formats provide news and in addition to *BBC News 24, Newsround, CNN, Al Jazeera* and *MTV* a number of other less traditional programmes were cited: *Have I Got News for You; Mock the Week; The Daily Show* (as well as *Radio 1* whose 'bite sized' human interest news appealed to several of our participants). It was important to know their broad levels of engagement with politics before beginning the session so we asked participants if they would vote in a general election. Just over half of our sample said they would vote (42/80); but almost as many said that either they would not vote (18/80) or that they were 'unsure' (20/80). When asked about which political party would receive their vote, there was some uncertainty. Indeed just 20 participants were sure that they would vote for a specific party (Labour 11/80; Liberal Democrats 5/80; Conservatives 4/80). The largest response, however, was simply in favour of the 'best person' and five participants said that they did not know who they would vote for but it would be 'Not Conservatives'. It was also important to have a sense of how politically engaged the participants believed they were prior to our discussion and we asked 'How Political Are You?' Their responses are as follows: 9 were 'very' interested in politics; 33 were 'a bit' interested; 24 were 'fairly' interested and just 14 said 'I am "Not" interested'.

Young people in the news media: Writing the news

We invited participants to work from photographs taken from the television news sample and to develop their own news stories, but with an eye to producing something that they might see on the news. This exercise allowed us to examine the extent to which certain phrases, images and themes that appear frequently in news coverage may be recalled and reproduced. Indeed it is striking that a number of recurring motifs featured in both our television news sample and in the invented news stories our participants produced. Thus, whether or not participants claimed to watch television news on a regular basis they were able to reproduce the themes, phrases and imagery that dominated our news reporting sample quickly and in some detail. It is worth noting that despite producing as many as 25 news stories in total when these are analysed their content relates to a remarkably narrow range of issues. Their reports are dominated by crime and different forms of anti-social behaviour. The story themes break down as follows with some stories covering more than one issue: Gun crime (n = 4); Date rape/spiked drink (n = 4); teenage crime/gangs (n = 2); stabbing (n = 4) binge drinking (n = 4). Other topics appear less frequently such as politicians' strategies to 'get youths off the streets' (n = 4); attracting the 'youth vote' (n = 1); Iraq war (n = 1) Olympic funding and cuts to youth projects (n = 1). It is worth reproducing some examples here to illustrate how accurate the tone is of the invented news stories:

Example of News Text: State School Students, Uxbridge

Another life lost as black teenager Marlon King was stabbed while playing football with his friends. This has had a huge effect on the boy's friends and family. They spoke of the growing racial tension in the area, Hackney. Gordon Brown has aimed to crack down on knife crime.

Example of News Text: Community Youth Group, NW London

Two gunned down at rap concert. Demonstrations outside parliament as protesters ask for music to be banned. MPs discuss new anti-gun crime measures.

Example of News Text: University Students, North London

Three youths were shot at a gig in Brixton. But the incident caused a public outcry, with parent and anti-gun groups protesting across the country. The public and media outcry caused the issue to be raised in parliament. Tougher laws were demanded to counteract this epidemic. On the day that the new 'zero tolerance to guns' bill was proposed Nick Robinson reported the story from Downing Street. Nick Robinson accused the Prime Minister of a whitewash as Gordon Brown had been keen to promote his new enterprising youth schemes because he wasn't doing enough to get the zero tolerance bill through parliament.

Example of News Text: Youth Group, West London

Gang related violence. Pressure on the government by protesters. Demonstrators campaign outside the local authorities' main building triggered by an incident that happened in a local nightclub where a fatal stabbing happened. A young person was stabbed to death. Local people view the incident as something that happens every day. Politicians simply don't care and do enough. [This] local youth worker has criticized the government and says that there's a lack of funding in facilities for young people. And the government just simply doesn't care. Meanwhile Gordon Brown visited a local school.

Example of News Text: Female Grammar School Students, Essex

Last week we witnessed the deeds of the black youth culture and got their take on how other people perceived them. We spoke to a group of young black people male and female and listened to how they bide their time in deprived areas of London, we also spoke to a youth worker called Dave. He told us that he thinks we should take youths' opinions and ideas into consideration in order to cut down crime, with giving them something to do. Three youths we spoke to explained that boredom leads to drug abuse and violence, [...] although they expressed their ideas in music like rap and mc'ing.

Example of News Text: Community Group, NW London

A youth was shot over drugs at 12.30 in a club in central London. [One witness] says that the man was shot and killed over money. Witnesses say they saw two black males in hoodies running from the scene of the crime. The girlfriend talks of her loss.

Young people in the news media: Criminals, hoodies, drunks and anti-social behaviour orders (ASBOs)

In light of the very negative themes apparent in the news stories above (and indeed from our own analysis of television news) we were interested in how young people themselves felt they were represented within the news media and asked the question: When young people appear on the news what do you usually see? Across all groups regardless of social class, gender or ethnicity the findings were remarkably similar. Young people believe that they feature disproportionately in negative stories or as a young woman summed up *'Everything that's not good'* (F1, Group 7, Friendship Group 1, Brighton). It is worth just listing a sample of their responses below:

Yobbish behaviour (Group 1); crime (Group 1); gun [crime]; vandalism, and murder, assault (Group 1). Stabbings (Group 2); violence and crime (Group 7); binge drinking (Group 7); drugs (Group 7); it's only bad things (Group 7); gun crime, drug crime, murders (Group 4); guns and stuff (Group 3); negativity, hoodies, stabbing of students (Group 4); it's just crime, ASBOs, whatever else (Group 6). Underage pregnancy

(Group 7); always negative stuff (Group 6); crime (Group 6); gangs, drugs (Group 9); guns, drugs and music (Group 9); binge drinking, vandalism, drug dealing, knives (Group 10); ODs, violence (Group 8); stabbings, shootings (Group 8); gun crime a lot at the moment (Group 8); drugs and teen pregnancy and diseases (Group 2); kids carrying guns (Group 2);drugs and crime (Group 12); hoodies this, hoodies that (Group 11).

A school student described that you rarely see a young person on the news unless it is framed in a negative way:

> You never hear a happy story about a young person doing something good in the community. It's all about like ASBOs, people causing trouble. (Male, Group 3, State School Students)

The more middle-class interviewees often made an implicit distinction between themselves and those young people who featured in television news. As one university student put it:

> When I think of what the media say as young people I don't group myself in with that group because it's so negative (laughs). So like you know it's either gang culture, drugs, binge drinking, ASBOs, and I'm not saying that you know it obviously doesn't affect me, because it does. (F4, Group 6, University Students, North London)

The consequences of such stereotyping do not only potentially impact on older people but also affect young people themselves who learn to fear their peers. For some of our participants the iconic image of a youth wearing a 'hoodie' symbolises their image of a criminal as the following exchange between grammar school students illustrates:

> *M1*: when you think of a stereotypical kind of person – if someone got mugged – then you think of a picture of the mugger it would be ...
>
> *M2*: a youth ...
>
> *M1*: quite a big person, like eighteen [years old] with a hood up, like a lurking drug addict somewhere, that's the stereotype. (Group 10, Male Grammar School Students)

The working-class respondents, however, could speak to the images of young people from a position of having direct experience of the kind of lives the stereotype both evokes and simplifies. As one participant in a community group living close to incidents of knife crime explained, 'The reason why people wear hoodies is because they don't feel safe'. Others spoke of local crime in their area and how measures to help control anti-social behaviour are not working because 'they see ASBOs a sort of medal' (Group 3, State School Students, Uxbridge).

We asked for suggestions about any positive stories that could act as a counterpoint to these powerful negative images but our participants struggled to think of any with the exception of celebratory stories of young children's achievements, exceptional educational results ('Those triplets going to Cambridge'), athletes or celebrities:

> You never turn onto a news story and it's something nice, a student got their degree! It doesn't happen. Students got their degrees today, but they're talking about how the examining board is not working right [...] We'll have to lower the grades! (Group 11, Friendship Group/ University Students, Colchester)

> You might get on the local news like a local child has got to the final of the international chess tournament or something. Or they're like going for the Olympics in 2012 for diving or something like that [...] The good stories about kids are usually the ones at the end of the news. It's like a nice sort of touch at the end. (M1, Group 7, Friendship Group, Brighton)

> I was watching on the news today, they were all talking about kids today because all the exams had come out [...] they looked at one school in North London – even though this place has gone 12 per cent up from last year of getting their GCSE grades, they still turned around and made it, 'oh, so how's it feel still being below average?' [...] they looked at the bad points. So straight away, 'oh, why are you under average?' (M5, Group 9, Young Offenders)

A positive story features then 'only if a specific person has talent like Theo the young guy the footballer. He's really young he's got so much talent. You have to have talent or else they don't really have no other stories' (F3, Group 5, Community Group, NW London).

In fact the views above are supported very strongly by evidence from our television sampling which found that where young people were referenced in positive ways it was indeed the case that these mainly centred on celebrities and in particular young sportspeople (specifically Theo Walcott who was chosen to join the England football squad).

As one female student explained, the problem lies in categorising young people as a homogenous distinct entity:

> I think it would be better if [journalists] didn't report on young people at all as a separate like specimen [...] we do everything that a lot of older people do, and a lot of older people do a lot of the things that we do. I spent a guy's 40th birthday with him, the whole weekend. And he does more drugs than I've ever seen anyone do in my life, and I've never even touched them [...] and me 21, the assumptions that will be made...if it was [a headline] 'someone took drugs that weekend' and there was a picture. (Group 8, Friendship Group 2, Brighton)

A school student in another group agreed:

> I think [journalists] generalise more with young people. They wouldn't like target someone like a middle class group of ladies and say 'Oh, everyone's like this'. But with young people they seem to just generalise across every single young person in the media. I think it's because we don't really get a say on anything. We don't really have the power to argue back whereas middle class ladies, they have the ability to argue back and then they also have the vote as well. Whereas us youngsters, we're not even allowed to vote yet and so we don't really have a say about what goes on. (F2, Group 3, State School Students, Uxbridge)

Part of the problem was also seen to be because these representations reflect very accurately the relative lack of power that younger people have in society:

> It's much more angled towards the older generation the news generally anyway. So it is about kind of the things that parents have to fear a lot of the time. (F2, Group 7, Friendship Group 1, Brighton)
>
> If you're sort of 30 and you don't have a family and you haven't had a high level education these kind of things don't affect you so the only way you're represented in the media is in a negative way [...] if you're outside the mould you're just not represented in a lot of ways. (F1, Group 6, University Students, North London)

In a West London youth group some members highlighted that although drugs and pregnancy issues in relation to young people may have slipped from the news agenda (teen pregnancy received very few references during our media sampling period) these issues continue to link young people with social policy failures:

> If [teenage pregnancy is on the news] it seems to be in the context of the government failing people. So they may have targets for child poverty that have not been reached about taking young people out of poverty. And the fact that government have made pledges on halving teenage pregnancy, but yet we're the worst in Europe, it's the worst country in the world to be a kid or whatever. You know all these kind of hyperbolic statements. It's either the young people ... there's a negative story about what they're doing, or it's the government's fault. [Murmurs of agreement] But there's rarely anything positive. (M3, Group 2, Youth Group, West London)

Members of this group won a local award for developing a successful community-based project but their achievement failed to generate significant interest from even the local press. Participants talked of their close proximity to recent stabbings and how in their own experience the media are far more interested in negative and sensational stories involving young people:

> we went to tell the [local] newspaper what we was doing, they were saying 'you know, it don't sell newspapers' [...] It's like, if you were to put [a headline] 'friendly neighbourhood teens clean up' and then 'ravenous [*sic*] gangs stab to death some little boy', people are going to pick up the person getting stabbed to death more – you see the people that stabbed like Kodjo [Yenga] down the road[1] [...] a lot of them'd be proud to say they're part of that gang so that they're left alone kind of thing. Do you know what I mean? I think it's all down to fear really. [It's] scaremongering and it's like playing on people's fear and like their perceptions and stuff like that. [...] it's like trying to make a movie out of real life basically. Trying to glamourise and dramatise what's happening [...] When you're young you want exciting things all the time. And you've got the media basically feeding off of that. (M4, Group 2, Youth Group, West London)

Some believed that the level of negative representation meant that young people have no significant incentive 'to do anything good' as these acts will generate 'a little two lines in the local paper. But I mean if you do something bad then you get all this publicity and all this big drama' (F3, Group 2). As another participant in the same group describes it 'It's the new image to be in a gang and walk around as thugs'. Furthermore, it was also suggested that perhaps these images had a self-fulfilling prophecy:

> [Young people] don't really come on the news for something that they've done well in school, it's like 'oh, there's been a gun crime' or 'there's been a knife crime' or something like that. And then people get scared of us. So obviously...obviously the people in these pictures are our age so *we get that image as well, so we get scared of them* [emphasis added] because they're portrayed like that by the media. (F2, Group 3, State School Students, Uxbridge)

For those participants in our young offender group there seemed little point in their experience in trying to bring about change as they are not supported by those in power (whom they see as corrupt and ineffectual). As one member of that group session describes here:

> You'll meet a lot of young people that will say 'I'd love to change things, I'd love to change this life'. Yeah? 'I'd love to do something good, something good for my community'. But every time they stand up to do something they get shot down. And it happens at such a young age. 12, 13, yeah. This is where they start seeing politics 'oh yeah, it's lies and deceit and this and all that'. Yet I can go get that with my local crew. Local people down the street. I don't expect it from the people who are running this country. (M5, Group 9, Young Offenders)

Politics: Part of everyday life?

Participants were also asked to give their first thoughts on hearing the word 'politics'. Although their answers were reasonably diverse (from 'decision-making' to 'war') in the main the answers reflected a fairly Westminster-based view.

In the House of Commons when they all just have a go at each other and don't get nothing done; Government; House of Commons; War, Tony Blair, They all sit in a room and bark at each other (Group 1) House of Commons; MPs, power and money (Group 7) Government where people make all the big decisions (Group 3) The War (Group 6) managing change; it's everything, war and religion (Group 11); elections, Tony Blair (Group 12); decision-making (Group 10).

Indeed Tony Blair was referenced more frequently than any other MP in their responses; however, that was probably due to the fact that his imminent resignation was discussed heavily across all sections of the media. As one participant put it, 'he's resigning and it's the biggest thing at the moment' (M2, Group 3).

However, some of our participants quite deliberately thought beyond Westminster. As a university student explains here:

> I feel quite strongly about the fact that politics shouldn't be pigeon-holed into Westminster. I think it's about everything. I think it's about like just having the right to express what you think, what you feel. And I think that especially sort of in recent years there's been so many different arenas for political expression and I think part of the problem with politics today is the fact that people think of politics and they think of men in suits and Westminster... (F3, Group 6)

Her friend agreed that politics needs to be linked to the everyday experience and recalled an interesting television programme during the last general election which attempted to do this:

> It was a guy going into a pub saying, you know, I don't follow politics, you know I don't believe in politics, I don't vote. And the other guy was like, well, you care about how much your beer costs. It's kind of that idea that although it's associated with this kind of central group, that group underpins and infiltrates all areas of society more so than people are willing to recognise. (F2, University Students, North London, Group 6)

Others believed that politics need not be so removed from people's lives: 'Everything's politics isn't it? it's things that need to be changed' (M6, University students, Colchester, Group 11). As one grammar school student describes, 'I think everyone has a political view they just don't really express it' and another added 'being part of a society means you have to have [political views] 'cause politics is everything'. In this context it is interesting to note that despite expressing very negative initial reactions to the term 'politics' (as one member said, 'It's shit – I don't believe in politics' (Group 9), this was a view that also resonated with those in our young offender group.

> *M4*: You can talk all day about politics, you see what I'm saying? It's throughout your life. You see it? But it comes off of the politics

stage and you just start talking about your own experiences rather than politics. *What politics really stands for* innit. You see what I'm saying?

M5: At the same time, politics is our own experience [...] Our politics is every day.

M4: So you might feel we're coming off the subject but we're not. We're talking politics right now basically innit. You see it? But it's not [Westminster] type of politics.

Politics as pantomime: The Parliament photograph

The photograph which depicted the House of Commons scene provoked widespread recognition and derision across the groups generating typical comments such as 'punch and judy' 'so much like pantomime', 'squabbling in Westminster' and as a school student put it, 'It keeps going forwards and backwards like tennis'. A young woman summed up her instant reactions:

> With the Parliament picture, I just kind of thought how Parliament can sometimes be a bit of a joke. Like it's just a big pantomime. Like people will watch it for entertainment value [politicians are] just always trying to get one up on each other and don't really address the issue. (F3, Group 8, Friendship Group 2, Brighton)

The links between popular entertainment, politics and citizenship are discussed elsewhere (see van Zoonen 2005) but within the groups several participants noted the similarities between bickering in Parliament and arguing in a television drama:

> It seems a bit pointless when they're talking about an argument between like two MPs, and someone's trying to make the other person look bad. I think it would be more interesting to know about what's actually happening. They like to make it into a little soap drama. [Laughter] 'he said this, he said that'. You're watching like really posh, upper class people arguing, and it's just not as interesting as *EastEnders*. [Laughter] because you always hear about how they're just trying to make the other party look bad and make theirs look better. It seems like it's not actually about what they're doing for the country. (M1, Group 3, State School Students, Uxbridge)

It was clear that this photograph embodied what many of our participants saw as the significant divide between government and people.

Indeed it is worth noting that this picture was chosen surprisingly infrequently in the photo exercise. Thus our group of female grammar students explained that they just could not see how to 'fit it in' with a story they wanted to tell, 'Writing about something with the House of Commons in it, you've got to write about something quite in depth and complicated. You don't usually get people talking about crime and stuff, like black youths and stuff like that' and another group said, 'It was quite difficult to choose the photo of the politics ...' (Group 5).

It is also interesting that Parliament and television news coverage of politics were considered to be entirely interlinked because 'it's all based on what happens in Parliament'. As another participant put it, '*The Politics Show* and *Newsnight* boil down to being very similar to parliament and Prime Ministers Questions because they are just two sides arguing [...] There's no completely independent TV channel' (M2, Group 8). The image sparked discussion of absences in representation in terms of younger faces and as one participant put it 'I've never seen a black man talking in the House of Commons or anything like that' (Group 12). The photograph also led to discussion of BBC's *Question Time*. As a member of a youth group in West London explained, 'when I see it I think it's just theatre really. It's just pantomime isn't it?' Others commented on how politicians refuse to answer questions with the result that 'you can never get a view on what they're saying' (Group 3). The Parliamentary Channel was associated with scenes where 'they're just shouting at each other'. However, a minority of our participants were more interested in politics and relished the opportunity to debate topics with their peers:

> I've been always very interested in politics, and have done since I was a child. And it's something that I've watched and I've read newspapers from a young age. And that's just because my mum was, you know. And it's something that I've inherited from her. And I have lots of friends who are equally interested and are now seeking careers in politics because that's exactly what, you know, the way they've been brought up as well. And they relish it on the TV and I have big debates in the pub about *Question Time*. (M3, Group 2, Youth Group, West London)

If the House of Commons and political debate is seen as theatre then politicians are frequently cast as characters and few of our participants could relate to them:

> *M2*: It's the characters in politics as well. It's like, someone like David Cameron, it's like he went to Eton. It's like 'what the fuck?'

F3: Yeah, you want someone you can relate to.
M2: What would he know about me and you? (Group 2, Youth Group, West London)

In more affluent groups as well, David Cameron was perceived as representing politicians who are lacking conviction, 'politicians like David Cameron, every single policy he has it's not like what he believes, it's what he thinks everyone will like – 'cause they're trying to represent everyone' (Group 10). There was a recognition that at least David Cameron was attempting to connect with younger audiences via new media, 'like iPod talks and like videos on Youtube. It's good [...] But I think it's not easy for all politicians to do it and sometimes he can come across a bit fake himself' (M1, Group 3).

Who do politicians represent?

But the problem of trying to represent 'everyone' was also seen as a problem across the political spectrum where all politicians are trying to represent as one participant summed up 'the median vote – like the vote of the most average person they can possibly think of'. Participants thought that as a consequence, the clear differences between parties have become 'blurred'. As one explains, 'gone are the days when, you know, the Labour Party represented working people, working classes and Conservatives represented the middle classes'. Those being targeted are considered to be 'white people, usually sort of middle class and people with families, that kind of thing, like there's a very much a sort of family agenda all the time' (F1, Group 6, University Students, North London). Another added, 'There's no sort of big difference I don't think any more. I think all the ideas have just sort of blended' (F2, Group 7, Friendship Group 1, Brighton). This apparent lack of choice is seen to have serious consequences for voting behaviour:

> I've always wanted to vote, *but you feel like you can't*. And in a way, because I think the parties have got so similar, you almost – unless you feel very strongly about a particular issue that one of the parties is very strong on – you almost can't make a decision. (emphasis added) (F3, Friendship Group 2, Group 8, Brighton)

We were interested in which groups our participants believed that politicians represented. The vast majority of participants saw them as mainly representing older people with families (in other words not those in our groups) or their political party (they have to 'toe the line' to progress) or as many of our respondents put it, they simply represent 'themselves'. The following quote comes from a member of the young offender group:

> Themselves. They don't represent us. Most people would say they represent elder people, or standing members of the community. Old people. I'm not saying old people, I'm saying 30 upwards, yeah, married life people, children. (M5, Group 9, Young Offender)

We followed this up by asking 'Do politicians represent you?' Again responses were very negative across the groups (with some notable exceptions). In one group in particular participants went as far as to suggest that '[politicians] are the evil people'; 'They represent themselves and what they want only', 'They represent their pounds and their dollars. They don't represent me'. This community group was composed of black and Asian participants aged between 16–24 years. They appeared to be more disillusioned than other participants (with the exception perhaps of the young offenders group) and this seemed to be linked to their anger with the government in terms of the war in Iraq and equally linked to their identity as young Muslims. As one young woman explains

> The English Government and the American Government made it really hard for Muslims living in this country and America because before any bombing happened we might as well say we were living in peace. But the minute September the 11th came, that's it – the world had their own view about Islam and now it's a religion that's evil basically to everybody else. (F3, Group 5, Community Group, NW London)

There was also a great deal of mistrust in terms of 'what we are told' by the media about events.[2] One participant questioned the 'true' motivations of Bin Laden and his attack on the Twin Towers. 'There must have been a good enough reason to do that? No one's really going to take innocent lives like that just for the fun of it. They're obviously going to do it because there's a reason. But we don't know the reason because we haven't been told'.

However, as noted earlier there was less of a disconnection between those in our more affluent groups with one saying: 'I don't think you'd take on a prime minister's job if you were just in it for the money [...]

I genuinely believe they do it to help people' (Group 11, Male, Friendship/ University Students, Essex). Others thought that local government was a different matter to central government as 'back home, you tend to know at least one of the candidates [...] I think with local politics you can have a good opinion. I think the policies are always a lot more clearer and more specific to your needs and your sort of fears'. Notably those in higher education also saw themselves as being represented by politicians:

> Yeah I think being middle class in full time education, I think you are definitely taken into consideration but I think as some of these pictures show I think you know there are people who definitely feel like they're not included. (M5, Group 11, Friendship/University Students, Colchester)

There was a recognition – amongst those who felt represented – that other groups in society were not because 'there's certain connotations which go with people that [politicians] don't want to represent and they sort of stay away from them' (Male, Group 11, Friendship Group/ University Students). This was considered to be because there was such an educational and social class divide:

> Some politicians are quite well educated I think and some of them went to like Eton, that's like an extreme example but [it's] really difficult for them to relate personally to some of the people in the pictures you know someone who's living in a single family and they dropped out of school really young and they've got involved in a drug culture or something like that – I don't see how they can be represented. (M2, Group 11, Friendship Group/University Students)

During their group session, the young offender participants spoke at length about the lack of representation, futility of the criminal system, lack of opportunities for rehabilitation and how for some people 'It's called criminal life but it's not necessarily criminal it's just a way of surviving for certain people' (Group 9). These participants were all taking part in prison education schemes but it was clear that in this group – in contrast with the views expressed above – they could see little of their identity in relation to those in government. As a result they felt they were not interested in 'politics' because 'it would interest us if they took our feelings into consideration and our opinions into consideration. But because they don't do that [...] you just tend to [think] "well I don't really want to know nothing about politics" innit'. There was significant class bitterness expressed in many of their comments:

> The majority of these fucking Government people and all that lot was brought up with a silver fucking spoon in their mouth. They haven't been on the street. They don't know what it's like on the street. I've been on the street since I was 15. I lived actually on the street and that. So I know what the street's all...Like most of us here we hang about on the street the majority of the time and that because there's fuck all to do, so we know what the streets life is like. They don't because they've never fucking been on the street. They've always...They've been brought up with a silver spoon in their mouth, they've gone all...they've had money to go to fucking colleges and all this bollocks. (M3, Group 9, Young Offenders)

The Iraq war: A lasting legacy of mistrust?

It is worth noting that despite many of our participants expressing little interest in politics, the war in Iraq was mentioned spontaneously in every group session. Partly this was in response to one of our news writing photographs which depicted a large crowd and was mistaken frequently for an anti-war protest (in fact the protest concerned the right to animal testing, discussed in Chapter 7). This photograph did provoke considerable debate amongst many of the participants.[3] The following exchange amongst friends in Brighton (Group 8) is revealing and even a little poignant in its sense of loss:

> *F1*: people did used to care. I think after years of just seeing that caring who you vote for got you nothing and didn't actually make the differences you expected...
> *F3*: Like the war's the perfect example.
> *F1*: Exactly.
> *F3*: Like that was...I think that was the biggest protest that's ever happened in Britain.
> *F1*: And they just didn't listen.
> *F3*: They didn't do anything.
> *F1*: And so people just think 'what's the point?' Like, that's the attitude I think.
> *F2*: It's completely out of our hands.
> *F1*: It's not that people don't care, but it's just like what is the point?
> *F2*: We can't do anything. (Group 8, Friendship Group 2, Brighton)

There was, however, some confusion about the stated aims of the war. Thus a school student asked why it seems that 'everyone hates Tony Blair' and receives a range of responses from his class mates:

M3: ...sending troops to Iraq can be a good thing or it can be a bad thing.
F2: It's a bad thing – all the amount of troops that have died from it.
M2: It's got nothing to do with us.
M1: And there was no oil there.
F2: What?
M3: That's what they had a war about.

Participants discussed taking part in the anti-war demonstrations, the meaning of the Hutton report and the death of Dr David Kelly. For some the Iraq war was taken to symbolise mistrust in the government and media to tell the public the whole story. Some saw the war as representing the decline of public trust in the Labour government. As discussed earlier Muslim participants talked of how hard it was to live in the UK in light of these events. In other words there was no evidence here that our participants were apathetic and there was significant anger expressed that the war had taken place against the public wishes and also that troops were still based in Iraq. Others have noted that the age group represented in our study were against the war in Iraq from the outset and that their protests were belittled by the mainstream news media (Cushion 2007).

Issues of comprehension and engagement

During the sessions it was apparent that there were some gaps in knowledge amongst our participants. For example, there was confusion over the differences between local council and government. In terms of wider international issues this was not noted only in younger groups or those from low-income backgrounds but equally amongst groups of students. As one university student put it, 'Things like Israel and Palestine, I just zone out when that comes out I don't really know the background story ...' (Male, Group 11). This taps into a serious problem whereby younger audiences are not being informed by television news. This lack of knowledge is linked heavily to lack of interest (as has been noted by Philo and Berry 2004 in relation to audience understandings of the Israel-Palestine conflict). Television news was seen as a medium that 'you need to follow' as a grammar school student put it 'cause they kind of expect their viewers to watch every single day'.

This can lead to problems as one participant describes

> If you followed that through when you were sort of a young teenager and then you moved onto the news then I think you would be ok, but I think if you maybe haven't watched the news since you were our age, then dive straight in, then you might struggle with a few of those long running disputes. (M4, Group 11, Friendship Group, Univ Students, Colchester)

This was a view that was echoed across the groups as a grammar school student explains, 'they assume knowledge which is actually lacking so people don't bother watching it cause sometimes they just don't understand. If you haven't kept up with what's going on then if you're watching something like Newsnight it's quite specific about a really complicated issue, you just switch it off' (M4, Group 10).

This idea that news relates stories that unfold over months or years and it is impossible to 'catch up' was identified across many different groups. One female participant described it as 'like a story – I've just opened a book in the middle. I have to get to the beginning. And in the everyday world when everyone's working or studying you don't have the time to do that' (F3, Friendship Group Brighton, Group 7). The fact the political education does not feature as a compulsory part of the school curriculum was discussed in most of the groups as one participant put it 'why not political education instead of religious education?' As it stands it is fairly easy to go through the educational system without learning about politics:

> I went to the school that Tony Blair's daughter went to, we didn't even learn about politics there. Then you go on to A-Levels and you have a choice, and I didn't pick it. And then I came to university and I did do a module in my first term, and I didn't like it because everyone else had done politics before and I didn't know what was going on. (F1, Group 7, Friendship Group 1, Brighton)

Many of our participants relied on their parents or other family members for political education. As one female grammar school student recalls,

> *F2*: I remember growing up and my grandparents saying to me to vote for I think Tories (laughs).

F3: Labour, not conservative?
F2: Are they different?
F3: Yeah! (Group 12, Female Grammar School Students, Essex)

Others, however, would rather not vote than make a decision that was based on nothing other than simply 'what their parents did'. The following quotation represents this view and it is worth noting that every member of this group agreed that this was the reason that they would never vote:

> I think politics should be taught maybe in school from a young age so that kids have got an understanding of it, and not just when they get to the age of 16 or 18, when they get a little slip through the door that says 'what do you vote for?' And then you ask your mum and dad, 'who do you vote for?' And they say, 'oh, well I've been voting for Labour, so you may as well do the same.' [Murmurs of agreement throughout] You know if it was taught from an early age we'd be able to make up our own mind about who we want to vote for and why we want to vote for these people. (Group 4)

Entertaining and informing?

The final part of this chapter explores views on the current provision of television news and reporting of politics. As one girl put it 'The news is just like any other programme, it has to entertain' (Group 4). Certainly for some of those younger participants there was a desire to close the gap between news reporters and their audiences. Thus some wanted the news to be more like the music channels with presenters dressed in 'jeans and tee shirt' (State School Students, Group 1, Hillingdon). Others felt there was a possibility of making it more representative by other means:

> I don't expect them to suddenly start wearing hoodies and stuff but generally if you look it's only like older white middle class men and you've got to think about why more females aren't getting involved and more ethnic groups aren't getting involved and why they're not having the opportunity to get involved. (M1, Group 3, State School Students, Uxbridge)

Other obvious inclusions seemed to be 'more [openly] gay news presenters' (Group 6) and 'younger people' (Group 3). In each of the groups the ways in which audiences were addressed in terms of mode of speech was highlighted as a problem and indicative of the gap between presenter and audience:

> *M4*: The way they dress and the way they speak is like...
> *F6*: Sophisticated.
> *M4*: ...it's like proper English, it's like 'and now moving onto the scene where the...' ...If they were talking like he does [puts on accent to impersonate M1] 'now we're like moving on, innit!' [Laughter]. (Group 1)

Others mentioned the lack of regional accents ('You never hear a scouse accent', Group 8) and described a typical news presenter as 'Monotoned information by a middle aged man'. As the grammar school students put it, there is a problem with 'the pronunciation thing as well like "La La La!"' (Group 10).

There were some ideas presented about making political news more participatory and indeed more accountable. In other words, programmes would be less about 'hearing what they're doing and never having your say' and this could be achieved by means of 'phone ins during the show, across a panel section' (Group 3) or using ten-minute slots for main news, debate, votes. Drawing on the new reality TV/Docu soap hybrid programmes, it was apparent that younger participants who are very much at ease with such formats wanted their politics presented by familiar popular faces (with the changes in broadcast culture documentaries tend to be fronted by familiar celebrity faces).

Of course, celebrities do know how to connect with their audiences and it may be that this desire for familiar presenters taps into the problems discussed earlier with politicians representing another world. In several groups however, the fly-on-the-wall documentary programme where Michael Portillo went to live with a single mother was highlighted as a useful example of how this could be achieved ('he lived with a single parent mother family and he realised he spent more a week on lunch than she had to live on' (Group 6). Ex soap actor Ross Kemp was also a popular suggestion as many participants watched his series on gangs around the world.

Although there was significant banter about how reporters speak and the gap between those on screen and the lived reality of the viewers lives, it is clear that as noted earlier younger participants (who called themselves, 'the MTV generation') were keen to have news that was more like a music channel with graffiti background, bright colours and reporters that were dressed informally. They also had clear ideas about how the news could be enlivened:

> Instead of having like a BBC London you'd have like...you could separate it into different parts. So it could be like elderly people, you could have 10–15 minutes talking about how their pension is going up or something within their community. So instead of BBC London you'd have like the London Pensioners or whatever. And you'd have like London Teenagers/Young Adults and everything – showing what's going on, what they can do with their money, like that. Instead of just for one person you could set it for a range of different people. And the people you would have [present it] would be lively people and they wouldn't be like dull, they wouldn't be like [in a dull, posh accent] 'and now we move on to the weather'. [Laughter] They could like pump it up for you and it'd be 'Now going onto Clare with the weather!' (M4, Group 1, State School Students, Hillingdon)

Others recalled programmes which seemed to blend comedy and political comment seamlessly. In particular one participant mentioned *Spitting Image* (ITV, 1984–1996) and mimicking phrases to entertain his family, 'Then just realising I had [made] a point on politics' (M5, Group 2, Youth Group, West London). Indeed comedy presenters such as Ricky Gervais and Bill Bailey were popular choices as presenters of political programmes.

However, at the same time there was a more fundamental shift required for some of our participants before significant change could take place. As the following exchange illustrates:

> *F3*: But I think there need to be changes like within the political class in order for that to happen really, because there's so much spin and stuff now that people don't...if you have interviewed a politician you don't believe what they are saying anyway. Or the general consensus is that they can't be trusted, so...
>
> *F2*: It just seems so detached now doesn't it (laughs)!
>
> *F3*: Yeah, it really does, and you don't...I don't feel like I can necessarily get a real grasp of what's going on *even though I'd like to.* (emphasis added) (University Students, North London Group 6)

As discussed earlier there was a clear desire for proper contextual knowledge about major issues. Thus, in terms of reporting the conflict in Iraq some of our participants wanted first to give audiences a 'history lesson':

> ...a kind of quite patronising but still history lesson and I'd start with the first time the British were in Iraq. I'd show the mustard gas, I'd show everything from the beginning and I would try and trace a line of the divide between the West and al Qaeda and I'd try and show things as the myth and as the propaganda that they are [...] maybe if there was just a bit more honesty and less propaganda people would see the whole thing. (F1, Group 6, University Students, North London)

Participants are keen to be given more relevant knowledge (e.g. concerning links between 'the clothes we buy and world poverty') but at the same time they want this to be delivered in a shorter time span. Existing programmes such as BBC *Newsround* were repeatedly praised for their ability to encapsulate the main issues in a condensed fashion. The challenge is to do so without patronising audiences and our participants were very aware of attempts by broadcasters and politicians to engage them.

Conclusion

The main aim of this chapter has been to shed light on how younger audiences relate to politics as represented in television news and current affairs. It seems clear that most of our participants expressed significant disillusionment with formal politics and this was expressed in different ways that related to their educational and social class background. The business of politics was closely associated with Westminster and considered to be 'punch and judy' style, very insular in nature and quite removed from the reality of their lives. At the same time, television news and current affairs were so linked into this arena that there was little to guide audiences in terms of analysis or being able to make an informed decision concerning voting. There was also a uniform response to the profile of young people in the news as being closely associated with crime and public disorder to the extent that this was seen to discourage positive behaviour.

Where news producers and presenters see the perceived crisis largely in terms of language and visuals – and certainly these aspects were raised as part of the problem – there are other more pressing and fundamental gaps in news provision. Many participants simply would not watch political coverage because it involved petty arguments between parties or mainly focused often on long-running issues that they knew little about. This does not suggest political apathy. Certain topics such as the continued presence of British troops in Iraq generated strong responses from those from quite different social backgrounds. At the same time the lack of action following protests would seem to underscore the futility of public action. Many participants expressed a desire for being taught about politics in the same way that other personal and social issues have become part of a school-based education. We might then surmise that the crisis can be tackled by taking some of this into consideration. While media values continue to prioritise elite sources and Westminster-based coverage, younger audiences remain excluded and many feel alienated. In short, the diversity of our participants' lives needs to become part of the mainstream news agenda.

Conclusion: Is Another Television News Paradigm Possible?

Television news is caught between and often combines two models of broadcasting: a patrician attachment to the formal institutions of 'democracy' with its formal, deferential 'stuffy' mode of address vs a more 'human interest', consumer orientated, celebrity focused and potentially (with BSkyB agitating for a change in broadcasting regulations) strongly editorialising news. These appear to be the choices open to viewers, choices shaped by the power of the state on the one hand and the market on the other. Some broadcasters think that to engage younger audiences, broadcasting needs to shift its centre of gravity from the patrician model to the market model (conceptualised as a shift from 'hard' news to 'soft' news). Yet while there is some evidence to suggest that young people have an interest in aspects of a soft news agenda, that evidence has to be interpreted carefully. For one thing, an attraction to a soft news agenda is shaped in a context in which the only other option is the austere patrician model. We should avoid falling into the trap of thinking that these are the only alternatives open to television news. The temptation to shift from hard news to soft would merely bring broadcasting even more into line with a world in which the language and culture of the market saturates everything. It would do little to empower audiences, inform audiences, give them access to a diverse range of voices and perspectives, problematise what is accepted as normal, challenge preconceived ideas and in general nourish critical faculties so that reason can prevail over some of the bizarre and irrational forces that currently shape political and economic life.

This is not to argue for the status quo, or that television news should not strive to be more 'popular'. Rather it is a call to uncouple what 'popular' might mean from the discourse of the market and its orientation towards competition, voyeurism and selling merchandise. For example, there are ways of doing what might on the surface appear to be 'light' entertainment stories that could open up a whole range of cultural politics that would be entirely appropriate to the lives of young people to address. A news outlet could cover the death of Michael Jackson in a way that simply recycles dominant platitudes, or in a way that digs into the cultural politics of this figure (especially around race). Even today in a multimedia environment, television is well placed to play a key role in providing the resources of information and debate essential for a democratic culture.

The reality, however, is that television news today is in crisis. The economic imperatives of the market (competition, profitability) are growing on television generally, even as advertising as a source of revenue is declining. This is a toxic combination that will exert ever-greater pressure on public service commitments such as news provision. In such a climate, risk, adventure and diversity, qualities that are desperately needed if news and current affairs provision is to re-engage audiences, will be in ever shorter supply. On the political front, television news finds itself locked into uncritical reportage of the institutions of political power that have a declining legitimacy amongst the wider public. Young people are at the cutting edge of this attitudinal trend vis-à-vis the core institution of representative democracy in the UK, Parliament and the political class that inhabits it. Thus the crisis around watching news about public life is connected with the crisis around public and political life itself. That there is a crisis is today appreciated by growing numbers of people. However, what most commentators cannot bring themselves to admit is that the roots of the crisis must be traced to the core social and economic relationships of the society we live in: namely capitalism, in its current neoliberal phase.

The limits of democracy within a framework of capitalist social relations set the limits of the possible in terms of informing and engaging the public with news programming. The ability of existing political institutions to engage citizens and to represent them has diminished as the power of capital has increased the unaccountability of economic and political structures. As consumer identities have expanded across the developed capitalist world, so citizen identities have diminished. Citizen identities are rooted in the sense that people have a stake in social and political affairs: a voice and means of expressing themselves to representatives. They are grounded in participation in political life and a sense that public opinion is not moulded from the top, by both media and political parties, but is forged by a genuinely diverse range of currents and experiences that can find expression in political institutions. We have seen that the expansion of consumer identities over citizen identities is intimately linked to the growth of the neoliberal model of capitalism, not least in the media industries themselves, and the consolidation of the Western political class, across parties, around a neoliberal agenda.

State and market structures are forging a relationship between politicians and the media that is characterised by a deep consensus on policy issues, leavened only by potentially dramatic conflict around individual politicians on narrowly conceived questions of competence and/or integrity. The absence of ideological divisions within the political class may be one explanation for the paucity of interrogation by the media of policy choices and policy-making, especially when the media are so reluctant to go outside the establishment for interpretive comment and analysis. The focus instead on personality and process over substantive questions around the 'ends' or aims of policy in turn feeds back into public perception of politics as a 'pantomime', a soap opera or remote spectacle.

Young people are at the cutting edge of this general alienation from mainstream politics and disengagement from mainstream news media. Being a young person has become progressively harder within neoliberal capitalism. As major beneficiaries of the welfare state in the post-war period, they are a particularly vulnerable sector of society as state provision and support is cut and replaced by social atomisation and individualisation of risk and opportunity. Employment, housing and education are all becoming profoundly problematic resources for young people to access without incurring huge debts. Working-class youth are of course more disadvantaged than middle-class youth and their alienation is correspondingly more severe. It is working-class youth who in the main figure on the television news as either victims or villains of crime (69 per cent of stories from our sample focusing on young people were crime related). Such a disproportionate emphasis on one topic must raise the suspicion that young people are being set up as the necessary folk devils of a law and order agenda pushed by television in concert with the tabloid media. In such representations we see the model of broadcasting tied subserviently to state power. It was not only young people who were pushed into crime-related stories. Many of the stories about ordinary people were also about law and order issues (41 per cent). However, when we extract celebrities out of the sample of young people and just focus on ordinary young people in the news, we find that a staggering 82 per cent of stories had a crime-related focus.

In the other predominant representation of young people (28 per cent of stories in our sample) as figures of sporting achievement, we can discern the consumer model of broadcasting, where celebrities and 'feel good' stories are the norm.

Only the BBC really has the resources and public service commitments to begin to address this problem. There have been attempts to change journalistic cultures, especially in radio news provision that has large younger audiences. There has been some cross-fertilisation of these approaches to television news, for example, where news stories are filtered through ordinary people's experiences and sometimes even their words and perspectives. But, welcome as this is, such developments are occasional and do not constitute the mainstream television news agenda, which remains fundamentally aligned to a conservative agenda in both content and style. In terms of including the public and young people specifically within news discourses as active and vocal participants across a range of subjects (and not just corralled into crime stories), our findings show how little progress has been achieved. Thus, television news reflects and reinforces the wider structures of political and social exclusion.

There seems little progress in addressing topics that are of particular interest to young people and which touch on their own experiences (for example, education, discrimination, substance abuse, personal finance). Similarly, there is little progress in including young people as interested parties in the coverage of mainstream topics, or exploring mainstream topics such as unemployment or housing specifically from a young person's angle. Broadcasters can always point to one or two single examples where they have done such things, but their routine journalistic practices point them in other directions, towards the elites and towards older people in the main. Even less progress has been made in exploring political themes *outside* Westminster politics that have attracted the interest and participation of young people (trade justice, human rights, the environment, animal welfare). Such political practices take place in a world that is largely unfamiliar to many political reporters who are trained to equate politics with what professional politicians are doing. Because direct action politics often calls the consensus politics that dominates the political mainstream into question, television journalists cannot engage with it and more often than not simply apply news frames that attempt to delegitimise it. In terms of including young people as interviewees (as participants, witnesses and commentators) and opening broadcasting up to them as producers, there is still a very long way to go.

The key to reversing trends towards disengagement and political alienation must include a change to a more inclusive kind of news-making that is less dominated by 'elite' actors and allows greater input by 'ordinary' people. For example, the public contributed only 9 per cent of the total speaking time for all news sources within our 517 politics stories. In contrast, politicians accounted for 62 per cent and 'experts' of various descriptions (academics, former officials, researchers, scientists, journalists, writers, doctors and lawyers) received 16 per cent of speaking time. Such statistics demonstrate the establishment orientation of the news agenda, where the public exists to be defined by an agenda set up elsewhere, rather than the public contributing to defining the news agenda.

We found these sharp skews towards the professional political class in terms of speaking opportunities reproduced across the range of other discussion based programmes outside the daily news broadcasts. Similar to the news reports, the political discussion programmes prioritised experts over the public, who (with the exception of *Question Time*) received only 8 per cent speaking time from our sample. While these programmes often gave in-depth consideration to topics that received only marginal attention in the news reports, they did so from a narrow range of perspectives that were no different from those dominating television news. In other words, their function could be said to extend and deepen the reach of the dominant ideology across a wider range of social issues. We continued to explore the relationship and overlaps between the news reports and the political discussion programmes around a specific issue: the anti-vivisection protests and campaigns that were in the news throughout May 2006.

Here we found that political activity outside the formal institutions of representative democracy is treated with great suspicion by the media, and that activists and protests are marginalised and even demonised. The critical issue for the media was not to engage in a debate over ends (is animal vivisection defensible and under what conditions?) but to focus relentlessly and sensationally on the selected means by some activists whose actions allow the media to mobilise a law and order agenda to defend established state and corporate practices and interests. This focus within the media underpinned very unequal access to the debate, strongly differential legitimation between participants, and with one or two exceptions, generally limited responsiveness (so that the claims of each side are generally not responded to by the other side). The media generally failed to exercise their capacity for independent investigative and editorial autonomy. In addition there was little space for television news agendas to be open to participant reframing. Nothing prevents television from playing a substantive role in contributing to the formation of public opinion based on the critical, interrogative interplay of debate; nothing except the social, political and economic interests television news is integrated into.

The concept of ideology has become progressively marginalised within media studies in recent years. Reflecting the diminishing capacity of academics to call social structures into question, the study of representations has progressively lost sight of the relationship between those representations and their roots in the social, political and economic dynamics of capitalism. Yet, as this book is being completed in 2009, it is precisely neoliberal capitalism that has entered into a severe and profound economic crisis. In the UK, huge amounts of public money has been spent propping up private sector banks, a bail out that will impact on the public purse for many years to come. The gap between the political class of Westminster and the public that this book has been analysing has now become mainstream news through the expenses scandal. This is a wider context that makes the question of ideology critique, which analyses the relationship between structures of representations and belief systems and their material roots in institutional and wider socio-economic and political conditions, so pressing. But from within the study itself, the empirical evidence generated calls for a return to explanatory methods that can make sense of the standardised, repetitive, narrow and hierarchical patterns in news provision that we have found, and their relationship to power and inequality. The concept of ideology ought to be a key component in such methods.

Given how poorly television news is serving young people, there may be a temptation by some to claim that television news' capacity to play a role in informing the public is dead, and that instead there will be a generational shift away from this old media to the new media of the Internet. In fact the available evidence for news consumption suggests that while television news as a source of news for young people has declined (and to the extent that it has, broadcasters perceive that there is a problem), it has hardly collapsed. It remains the main primary source of news for young people as it does for people in general. In the medium term at least, there is no evidence that the Internet will eclipse television as *the* main source of news for young people. What is already happening is that television news augments its broadcast services with web pages that offer non-linear access to news. This in turn could expand the possibilities for participation in public debate. However, if the Internet is seen as only a technological fix, simply another way of delivering the same content to audiences, then it will fail to enhance the attractiveness of television news to a broader demographic. Similarly, invitations to 'have your voice' heard lose their potential to engage people in a more participatory and interactive way when there is little sense that anyone within news organisations have a responsibility to respond or are in any way accountable to audiences. Currently, the predominant way that the Internet is disseminating news to young people is probably in the form of headline news items that they come across while engaged in other Internet activities, such as logging onto their email. The reduction of news to brief headlines or updates on rolling news items (whether on the Internet or via text alerts to mobile phones) cannot be considered a great advance in terms of fostering an informed and engaged citizenship.

To go beyond such technological fixes requires drawing on research into audiences that goes beyond 'market research', with its narrow focus on identifying negative and positive responses to stimuli. We need to situate audiences as social beings in a particular society at a particular time to properly understand how they consume television news, or why they do not consume it. Our research into how young people from different class backgrounds and ethnicities engage with news about politics suggests that profound alienation rather than apathy informs their attitudes to both television news and Westminster party politics. Together, television news and the mainstream political parties have constructed a soap opera called 'Westminsterland', a spectacle with a declining fan base. This was especially true amongst working class and ethnic minority audiences, but even the middle-class respondents reported a strong sense of distance and remoteness vis-à-vis the centres of political power, as well as frustration with and scepticism about television's coverage of politics. Again and again participants demonstrated an awareness that seems to be beyond television journalists: namely that there is a life of politics outside Westminster and 'everyday politics' is something that people are naturally curious and interested in. In addition to the conflation between politics and Westminster which television routinely makes, the formality of dress and behavioural codes, the language used by journalists, the knowledge they assume on the part of the audience and the absence of contextual explanation for news stories, the proximity of news to power, the tedious conventionality of news were just some of the other problems our young interviewees found with television news.

The overall impression is that both the political establishment and television news, both of which are seen to be closely related to one another, have profound problems of legitimacy in the eyes of working-class and middle-class groups, and profound problems in reaching such groups and engaging with them. But the bigger question for us is that society itself has a profound problem when the institutions of representative democracy and organs responsible for constructing a public sphere of rational debate are perceived to be demonstrably failing to represent the public in both its diversity *and* its inequalities. It would be intellectually dishonest to suggest that the audience responses to news and politics are merely a problem of 'perception' (this is the kind of excuse politicians seek refuge in when their popularity is on the decline). These 'perceptions' are not irrational or ungrounded. In fact young people are drawing a reasonable conclusion based on their experience of both politics and its mediation by television news. And therein lies a seed of hope.

Notes

4 Content analysis of television news

1. A *'principal/main/subsidiary'* or '1/M/S' broad category match occurred when for the same story one coder recorded a story subject (e.g. politics) or 'actor' (e.g. politicians) as *'principal'* or *'main'* and the second coder recorded the same topic/actor as either *'principal'*, *'main'* or *'subsidiary'*. Within the reliability sample of 100 stories, coders reached agreement on 171 of 215 '1/M/S' story subjects (80 per cent) and 146 of 155 *'main/subsidiary'* (M/S) story actors (94 per cent). Reliability testing showed exact matches for 223 of 245 news sources (91 per cent). The sample was analysed by one coder employed to work on the project full time, and a second coder who was trained and employed over a three-month period. When common disagreements were identified, they were discussed in group consultation and related stories were then checked and recoded according to the agreed method.
2. Analysis of broad topics is, of course, an imperfect method of measuring relative levels of 'soft' and 'hard' news. Not all international news stories in our sample were, for instance, concentrated on serious issues, and crime news sometimes included serious analysis of law and order policy and practice in the UK. Indeed some scholars define 'soft news' more qualitatively by story framing, such that much contemporary reporting on politics could be described as 'soft' news since it focuses on personalities and 'hoopla' rather than policies (e.g. Patterson 2000). With this important caveat in mind, the distinction between 'hard' and 'soft' news in conjunction with these broad topics does provide a useful if preliminary picture of the news agenda. In the case of international news stories in our sample, the most common topics were confrontation with Iran over its nuclear programme (30 stories), an earthquake in Indonesia (31), and conflicts in Iraq (127), Afghanistan (29), Israel-Palestine (18) and Darfur (16). Similarly, most of the big crime stories in our sample related to violent and sensational cases, with particular focus on knife crime after a number of fatal knife attacks (which we discuss in Chapter 5).
3. The fact that *Newsnight* and *Channel 4 News* which were longer in duration and which featured many more interviews (and thus more overall speaking time by 'news sources') was thus the principal reason that politicians and not the public were the main quoted group across the full sample of seven programmes.
4. Almost all of the stories about human rights related to a backlash against Britain's Human Rights Act, which was accused of favouring the rights of criminals above those of their victims.
5. Again the length of *Newsnight* and *Channel 4 News* have distorted the overall averages somewhat; yet if we look at the median length of speaking time instead of the average, we still find considerably longer speaking opportunities for politicians (17 seconds) and 'experts' (18 seconds) than for the public (10 seconds).

5 The symbolic criminalisation of young people

1. This figure of 96 domestic stories excludes 1 item about YP and violent crime in the US. The figure of 127 domestic crime stories excludes 7 stories about YP and crime internationally.
2. This analysis includes all stories involving YP and violent crime in our sample, almost all of which were domestic stories.
3. The victim's ethnicity was identified in contemporaneous newspaper and online news sources.
4. Includes one story about knife crime in the US.
5. According to *ITV News*, ITV1 22.30, 30 May, 2006.

6 The monopolisation of political discourse

1. A version of the same analysis was previously published as an article in *Television and New Media*. The present analysis differs through the inclusion of *Newsround* and also via the inclusion of material relating to political discussion programmes. The inclusion of *Newsround* in the samples of news stories means that the statistical results provided here vary from those given in the *Television and New Media* article.

7 The boundaries of political debate: Animal rights

1. *The Independent*, May 20, Jackie Ballard, p. 38.
2. *The Daily Telegraph*, 29 May 2006, 'Public Turns on Animal Terrorists' by Philip Johnston, p. 1.
3. *The Sunday Telegraph*, 14 May 2006, 'At Last, the City Shows Some Moral Fibre' by Sylvia Pfeifer, p. 3.
4. *The Daily Mail*, 27 May 2006, 'Oxford Wins New Block on Animal Rights Protesters' by Olinka Koster, p. 35.

9 Talk back: Young audiences and reception

1. Kodjo Yenga was a 16-year-old A-level student from Hammersmith, West London who was stabbed and killed on 14 March 2007 after being lured to a fight and then ambushed by a gang of teenagers.
2. The news was considered to be biased to the extent that it was misreporting events such as the car bomb found outside the Tiger Tiger nightclub in London (two days after Gordon Brown became Prime Minister). 'It's like the bomb that they found two days ago in central London, the first thing they said – "oh, is this linked to terrorists?" They don't know that. It could have been anyone that done it. It could have been someone that's got hatred for the owner of that club. It could have been anyone' (M1, Group 5, Community Group, NW London).
3. This photograph was also used as a 'hook' for a 'people's protest' news story, 'Today's a big day in politics. The public for the past five years have rallied together in their outrage and distrust of the Government. Scenes of protest are common in the streets of England. And Parliament itself has been split

by the Prime Minister's decision to back the US in their War on Terror and the war in Iraq. Our own correspondent, Garth Harangay [is] outside number 10 Downing Street where he's waiting to speak with our new Prime Minister, who now has to handle the consequences of his predecessor's decisions. How will he implement his plans on British politics? And where will that leave Great Britain?'(Group 8, Friendship Group 2, Brighton).

Bibliography

Amin, S. (1997) *Capitalism in the Age of Globalization: The Management of Contemporary Society*, London: Zed Books.

Barnett, S. (2002) 'Will a Crisis in Journalism Provoke a Crisis in Democracy?', *Political Quarterly*, 73: 400–8.

Barnett, S., Seymour E. and Gabor I. (2000) *From Callaghan to Kosovo: Changing Trends in British Television News 1975–1999*, London: ITC.

Baum, M. A. (2003) 'Soft News and Political Knowledge: Evidence of Absence or Absence of Evidence?', *Political Communication*, 20: 173–90.

Behr, R. L. and Iyengar, S. (1985) 'Television News, Real-World Cues, and Changes in the Public Agenda', *The Public Opinion Quarterly*, 49 (1): 38–57.

Benjamin, W. (1999) *The Arcades Project*, translated by Howard Eiland and Kevin McLaughlin, Harvard: Harvard University Press.

Bennett, Stephen (1997) 'Why Young Americans Hate Politics and What We Should Do about It', *Political Science and Politics*, 30 (1): 47–53.

Bennett, Stephen (1998) 'Young Americans' Indifference to Media Coverage of Public Affairs', *Political Science and Politics*, 31 (3): 535–41.

Bennett, W. L. (1990) 'Toward a Theory of Press-State Relations in the United States', *Journal of Communication*, 40 (2): 103–25.

Bennett, W. Lance (2007) 'Relief in Hard Times: A Defense of Jon Stewart's Comedy in an Age of Cynicism', *Critical Studies in Media Communication*, 24 (3): 278–83.

Bennett, W. L., Pickard, V. W., Iozzi, D. P., Schroeder, C., Lagos, T. and Caswell, C. E. (2004) 'Managing the Public Sphere: Journalistic Construction of the Great Globalization Debate', *Journal of Communication*, 54 (3): 437–55.Bentley, T. and Oakley, K. (1999) *The Real Deal, What Young People Really Think about Government, Politics and Social Exclusion*, London: DEMOS.

Benton, T. and Redfearn, S. (1996) 'The Politics of Animal Rights: Where Is the Left?', *New Left Review 1*, (215).

Berry, D. and Theobald, J. (eds) (2006) *Radical Mass Media Criticism, A Cultural Genealogy*, Montreal: Black Rose Books.

Besley, J. C. (2006) 'The Role of Entertainment Television and Its Interactions with Individual Values in Explaining Political Participation', *International Journal of Press/Politics*, 11 (2): 41–63.

Bevir, M. and Rhodes, R. A. W. (2006) *Prime Ministers, Presidentialism and Westminster Smokescreens Political Studies*, 54 (4): 671–90.

Bhavnani, K. K. (1991) *Talking Politics: A Psychological Framing for Views from Youth in Britain*, Cambridge: Cambridge University Press.

Blair, T. (2007) 'Tony Blair's "Media" Speech: The Prime Minister's Reuters Speech on Public Life', *Political Quarterly*, 78 (4): 476–87.

Blumler, J. and Gurevitch, M. (1995) *The Crisis of Public Communication*, London: Routledge.

Brants, K. (1998) 'Who's Afraid of Infotainment?', *European Journal of Communication*, 13 (3): 315–35.

Brewer, P. R. and Sigelman, L. (2002) 'Political Scientists as Color Commentators: Framing and Expert Commentary in Media Campaign Coverage', *Press/Politics,* 7 (1): 23–35.

Briggs, A. (1995) *The History of Broadcasting in the United Kingdom: Volume 5: Competition*, Oxford: Oxford University Press.

Brookes, R., Lewis, J. and Wahl-Jorgensen, K. (2004) 'The Media Representation of Public Opinion: British Television News Coverage of the 2001 General Election', *Media, Culture & Society*, 26 (62): 63–80.

Buckingham, David (2000a) *The Making of Citizens: Young People, News and Politics*, London: Routledge.

Buckingham, D. (2000b) *After the Death of Childhood, Growing Up in the Age of Electronic Media*, Cambridge: Polity.

Bunting, M. (2003) 'A Conversation Conducted in Two Different Languages', *Guardian*, 29 November, p. 26.

Cappella, J. N. and Jamieson, K. H. (1996) 'News Frames, Political Cynicism, and Media Cynicism', *The Annals of the American Academy* (546): 71–84.

Cappella, J. and Jamieson, K. H. (1997) *Spiral of Cynicism: The Press and the Public Good*, Oxford: Oxford University Press.

Caragee, K. M. and Roefs, W. (2004) 'The Neglect of Power in Recent Framing Research', *Journal of Communication*, 54 (2): 214–33.

Carter, C., Messenger Davies, M., Allan, S. and Mendes, K. (2009) *What Do Children Want from the BBC? Children's Content and Participatory Environments in an Age of Citizen Media*, Cardiff: Cardiff School of Journalism, Media and Cultural Studies.

Carter, S. and Henderson, L. (2005) 'Approaches to Qualitative Data Collection in Social Science', in A. Bowling and S. Ebrahim (eds), *Handbook of Health Research Methods: Investigation Measurement and Analysis,* Milton Keynes: Open University Press.

Chibnall, S. (1977) *Law and Order News: An Analysis of Crime Reporting in the British Press*, London: Tavistock Publications.

Cohen, Stanley (2002, third edition) *Folk Devils and Moral Panics*, London: Routledge.

Coleman, C. and Moynihan, J. (2003) *Understanding Crime Data*, Maidenhead: Open University Press.

Connell, R. W. (1971) *The Child's Construction of Politics*, Melbourne: Melbourne University Press.

Cottle, S. and Rai, M. (2006) 'Between Display and Deliberation: Analyzing TV News as Communicative Architecture', *Media, Culture & Society,* 28 (2): 163–89.

Curran, J. and Seaton, J. (2003) *Power without Responsibility: The Press, Broadcasting and New Media in Britain*, 6th edition, London: Routledge.

Cushion, S. (2007) 'Protesting Their Apathy? An Analysis of British Press Coverage of Young Anti-Iraq War Protestors', *Journal of Youth Studies*, 10 (4): 419–437.

Cushion, S. (2009) 'Discouraging Citizenship? Young People's Reactions to News Media Coverage of Anti-Iraq War Protesting in the UK', *Young*, 17 (2): 123–43.

Dardis, F. E. (2006) 'Marginalization Devices in U.S. Press Coverage of Iraq War Protest: A Content Analysis', *Mass Communication and Society*, 9 (2): 117–35.

Deacon, D., Wring, D., Billig, M., Downey, J., Golding, P. and Davidson S. (2005) 'Reporting the 2005 U.K. General Election'. www.electoralcommission.org.uk (accessed 12 February 2007).

Deardourff, J. (1996) 'Coverage Drives Campaigns: Reflections on the Decline of Political Reporting', *Press/Politics,* 1 (2): 100–10.

Dixon, T. L. and Azocar, C. (2007) 'Priming Crime and Activating Blackness: Understanding the Psychological Impact of the Overrepresentation of Blacks as Lawbreakers on Television News', *Journal of Communication,* 57 (2): 229–53.

Dixon, T. L. and Linz, D. (2000) 'Overrepresentation and Under-representation of African Americans and Latinos as Lawbreakers on Television News', *Journal of Communication,* 50 (2): 131–54.

Edwards, D. and Cromwell, D. (2006) *Guardians of Power: The Myth of the Liberal Media,* London: Pluto Press.

Edwards, G. (2004) 'Habermas and Social Movements: What's "New"?' in Nick Crossley and John Michael Roberts (eds), *After Habermas, New Perspectives on the Public Sphere,* Oxford: Blackwell.

Eilders, C., Degenhardt, K., Herrmann, P. and Der Lippe, M. V. (2004) 'Surfing the Tide: An Analysis of Party and Issue Coverage in the National Election Campaign 2002', *German Politics,* 13 (2): 218–42.

Electoral Commission (2001) *Election 2001: The Official Results.* http://www.electoralcommission.org.uk/elections/2001report.cfm (accessed 19 March 2007).

Eliasoph, N. (1998) *Avoiding Politics: How Americans Produce Apathy in Everyday Life,* Cambridge: Cambridge University Press.

Entman, R. (1993) 'Framing: Toward Clarification of a Fractured Paradigm', *Journal of Communication,* 43 (4): 51–8.

Ericson, R., Baranek, P. M. and Chan, J. (1991) *Representing Order: Crime, Law and Justice in the Mass Media,* Maidenhead: Open University Press.

Fahmy, Eldin (2008) 'Tackling Youth Exclusion in the UK: Challenges for Current Policy and Practice', *Social Work and Society,* 6 (2). http://www.socwork.net/2008/2/special_issue/fahmy/index_html/?searchterm=None (accessed April 2009).

Fairclough, N. (2000) *New Labour, New Language?* London: Routledge.

Farnsworth, S. J. and Lichter, S. R. (2004) 'Increasing Candidate-Centered Television Discourse: Evaluating Local News Coverage of Campaign 2000', *Press/Politics,* 9 (2): 76–93.

Franklin, B. (1997) *Newszak and News Media,* London: Edward Arnold.

Franklin, B. (2004) *Packaging Politics: Political Communications in Britain's Media Democracy,* 2nd edition, London: Edward Arnold.

Franklin, B. and Petley, J. (1996) 'Killing the Age of Innocence: Newspaper Reporting of the Death of James Bulger', in Jane Pilcher and Stephen Wagg (eds), *Thatcher's Children, Politic: Childhood and Society in the 1980s and 1990s,* London: Falmer Press.

Foley, M. (2004) 'Presidential Attribution as an Agency of Prime Ministerial Critique in a Parliamentary Democracy: The Case of Tony Blair', *British Journal of Politics & International Relations,* 6 (3): 292–311.

Foot, P. (2005) *The Vote: How It Was Won and How It Was Undermined,* London: Viking.

Furlong, A. and Cartmel, F. (1997) *Young People and Social Change: Individualisation and Risk in Late Modernity*, Buckingham: Open University Press.
Gan, F., Teo, J. L. and Detenber, B. H. (2005) 'Framing the Battle for the White House: A Comparison of Two National Newspapers' Coverage of the 2000 United States Presidential Election', *Gazette: The International Journal for Communication Studies*, 67 (5): 441–67.
Gerbner, G. (2003) 'Television Violence: At a Time of Turmoil and Terror', in Gail Dines and Lean Humez (eds), *Gender, Race and Class in Media: A Text-Reader*, London: Sage.
Gerbner, G., Gross, L., Morgan, M. and Signorielli, N. (1986) 'Living with Television: The Dynamics of the Cultivation Process' in Jennings Bryant and Dolf Zillmann (eds), *Perspectives on Media Effects*, London: Lawrence Erlbaum Associates.
Giddens, A. (1991) *Modernity and Self-Identity: Self and Society in the Late Modern Age*, Cambridge: Polity Press.
Giddens, A. (1998) *The Third Way: The Renewal of Social Democracy*, Cambridge: Polity Press.
Giddy, P. (2005) 'Just Like Watching Brazil', *Red Pepper*, October: 24–5.
Gilliam, D. F. and Iyengar, S. (2000) 'Prime Suspects: The Influence of Local Television News on the Viewing Public', *American Journal of Political Science*, 44 (3): 560–73.
Giroux, H. A. (2003) *The Abandoned Generation: Democracy beyond Fear*, Basingstoke: Palgrave.
Gitlin, T. (1980) *The Whole World Is Watching*, London: University of California Press.
Glasgow Media Group (1976) *Bad News*, Vol. 1, London: Routledge and Kegan Paul.
Glasgow Media Group (1980) *Bad News* Vol. 2, *More Bad News*, London: Routledge and Kegan Paul.
Goddard, P., Scammell M. and Semetko, H. M. (1998) 'Too Much of a Good Thing? Television in the 1997 Election Campaign', in I. Crewe, B. Gosschalk and J. Bartle (eds), *Political Communications: Why Labour Won the General Election of 1997*, London: Frank Cass.
Golden, F. (1985) 'Here Come the Microkids', in T. Forester (ed.), *The Information Technology Revolution*, Oxford: Basil Blackwell.
Goode, E. and Ben-Yehuda, N. (1994) *Moral Panics: The Social Construction of Deviance*, Oxford: Blackwell.
Goodman, Steven (2003) *Teaching Youth Media: A Critical Guide to Literacy, Video Production, and Social Change*, New York: Teachers College Press.
Graber, Doris (1980) *Crime News and the Public*, New York: Praeger.
Graber, D. A. (1993) 'Failures in News Transmissions: Reasons and Remedies' in Philip Gaunt (ed.), *Beyond Agendas, New Directions in Communication Research*, Westport, CT: Greenwood Press.
Graber, D. A. (2003) 'The Rocky Road to New Paradigms: Modernizing News and Citizenship Standards', *Political Communication*, 20 (2): 145–8.
Griffin, C. (2004) 'Representations of the Young', in Jeremy Roche, Stanley Tucker, Rachel Thomson and Ronny Flynn (eds), *Youth in Society*, London: Sage.
Habermas, J. (1987) *The Theory of Communicative Action, The Critique of Functionalist Reason*, Vol. 2, Cambridge: Polity Press.

Habermas, J. (1996) *The Structural Transformation of the Public Sphere*, Cambridge: Polity Press.
Hall, T. and Williamson, H. (1999) *Citizenship and Community*, Leicester: Youth Work Press.
Hall, T., Williamson, H. and Coffey, A. (2000) 'Young People, Citizenship and the Third Way: A Role for the Youth Service?' *Journal of Youth Studies*, 3 (4): 461–72.
Hargreaves, I. and Thomas, J. (2002) *New News, Old News, An ITC and BSC Research Publication*, ITC/BSC.
Harvey, S. (2006) 'OFCOM's First Year and Neoliberalism's Blind Spot: Attacking the Culture of Production', *Screen*, 47 (1): 91–105.
Hebdige, D. (1979) *Subculture: The Meaning of Style*, London: Methuen.
Heffernan, R. (2003) 'Prime Ministerial Predominance? Core Executive Politics in the UK', *British Journal of Politics & International Relations*, 5 (3): 347–72.
Henderson, L. (2007) *Social Issues in Television Fiction*, Edinburgh: Edinburgh University Press.
Henn, M. and Weinstein, M. (2006) 'Young People and Political (In)activism: Why Don't Young People Vote?', *The Policy Press*, 34 (3): 517–34.
Hobsbawm, E. (1987) *The Age of Empire*, London: Weidenfeld and Nicolson.
Home Office Standing Conference on Crime Prevention (1989) *Report of the Working Group on the Fear of Crime*, London: Home Office.
Hooghe, Marc (2002) 'Watching Television and Civic Engagement: Disentangling the Effects of Time, Programs, and Stations', *Press/Politics*, 7 (2): 84–104.
Inglehart, R. (1997) *Modernization and Postmodernization. Cultural, Economic and Political Change in 43 Societies*, Princeton, NJ: Princeton University Press.
Iyengar, S. (1991) *Is Anyone Responsible? How Television Frames Political Issues*, Chicago, IL: University of Chicago Press.
Jasper, J. M. and Poulsen, J. (1993) 'Fighting Back: Vulnerabilities, Blunders, and Countermobilization by the Targets in Three Animal Rights Campaigns', *Sociological Forum*, 8 (4).
Jewkes, Y. (2004) *Media and Crime*, London: Sage.
John, D. R. (2002) 'Consumer Socialization of Children: A Retrospective Look at Twenty-Five Years of Research' in Flemming Hansen, Jeanette Rasmussen and Anne Martensen (eds), *Children: Consumption, Advertising & Media*, Copenhagen: Copenhagen Business School.
Kateb, George (1981) 'The Moral Distinctiveness of Representative Democracy', *Ethics*, 91 (3): 357–74.
Kellner, Douglas (1990) *Television and the Crisis of Democracy*, San Francisco, CA: Westview Press.
Keen, A. (2008, second edition) *The Cult of the Amateur: How Blogs, MySpace, YouTube and the Rest of Today's User-Generated Media Are Killing Our Culture and Economy*, London: Nicholas Brealey Publishing.
Kerbel, M., Apee, S. and Ross, M. H. (2000) 'PBS Ain't So Different: Public Broadcasting, Election Frames and Democratic Empowerment', *Journal of Press/Politics*, 5 (4): 8–32.
Kevill, S. (2002) *Beyond the Soundbite*, London: BBC.
Kimberlee, R. (2002) 'Why Don't British Young People Vote at General Elections?', *Journal of Youth Studies*, 5 (1): 85–98.
Klein, N. (2000) *No Logo*, London: Harper Collins.

Klite, P., Bardwell, R. and Salzman, J. (1997) 'Local TV News: Getting Away with Murder', *Press/Politics*, 2 (2): 102–12.
Lewis, J., Inthorn, S. and Wahl-Jorgensen, K. (2005) *Citizens or Consumers?* Maidenhead: Open University Press.
Leys, C. (2001) *Market-Driven Politics, Neoliberal Democracy and the Public Interest*, London: Verso.
Lichter, R. (2001) 'A Plague on Both Parties, Substance and Fairness in TV Election News', *Journal of Press/Politics*, 6 (3): 8–30.
Lichter, R. and Smith, T. (1996) 'Why Elections Are Bad News, Media and Candidate Discourse in the 1996 Presidential Primaries', *Journal of Press/Politics*, 1 (4): 15–35.
Lipscultz, J. H. (2007) 'Framing Terror; Violence, Social Conflict, and the "War on Terror"', *Electronic News*, 1 (1): 21–35.
Livingstone, S. (2007) 'Interactivity and Participation on the Internet: Young People's Response to the Civic Sphere', in P. Dahlgren (ed.), *Young Citizens and New Media: Learning for Democratic Participation*, London: Routledge.
Livingstone, S. (2009) *Children and the Internet*, Cambridge: Polity.
Livingstone, S., Couldry, N. and Markham, T. (2007) 'Youthful Steps towards Civic Participation: Does the Internet Help?', in B. Loader (ed.), *Young Citizens in the Digital Age: Political Engagement, Young People and New Media*, London: Routledge.
Lloyd, J. (2002) 'Media Manifesto', *Prospect*, October: 48–53.
Lloyd, John (2004) *What the Media Are Doing to Our Politics*, London: Constable.
Maguire, M. (2002) 'Crime Statistics: The Data Explosion', in Mike Maguire, Rod Morgan and Robert Reiner (eds), *Oxford Handbook of Criminology*, Oxford: Oxford University Press.
Mair, P. (1997) *Party System Change: Approaches and Interpretations*. Oxford: Clarendon Press.
Manning, B. (1999) *The Far Left in the English Civil War*, London: Bookmarks.
Marsh, D., Richards, D. and Smith, M. (2003) 'Unequal Plurality: Towards an Asymmetric Power Model of British Politics', *Government and Opposition*, 38 (3): 306–32.
McFarlane, T. and Hay, I. (2003) 'The Battle for Seattle: Protest and Popular Geopolitics in the Australian Newspaper', *Political Geography*, 22 (2): 211–32.
McGuigan, J. (1992) *Cultural Populism*, London: Routledge.
McKay, G. (1998) *DiY Culture, Party & Protest in Nineties Britain*, London: Verso.
McLeod, D. M. and Detenber, B. H. (1999) 'Framing Effects of Television News Coverage of Social Protest', *Journal of Communication*, 49 (3): 3–23.
McNair, B., Hibberd, M. and Schlesinger, P. (2003) *Mediated Access, Broadcasting and Democratic Participation in the Age of Mediated Politics*, Luton: University of Luton Press.
Miller, D. (2003) *Tell Me Lies: Propaganda and Media Distortion in the Attack on Iraq*, London: Pluto Press.
Miller, W., Timpson, A. M. and Lessnoff, M. H. (1996) *Political Culture in Contemporary Britain: People and Politicians. Principles and Practice*, Oxford: Clarendon Press.
Mishin, A. (1975) *Bourgeois Democracy in Theory and Practice*, Moscow: Novosti Press Agency.

Mitchell, A. and Sikka, P. (2006) 'On Pensions and Poverty', *Socialist Worker* December, no. 2031. http://www.socialistworker.co.uk/art.php?id=10320 (accessed April 2009).

Mizen, P. (2004) *The Changing State of Youth*, Basingstoke: Palgrave MacMillan.

Mosco, V. (2005) *The Digital Sublime: Myth, Power and Cyberspace*, Cambridge, MA: MIT Press.

Munday, M. (2007) 'The Myth of the My Space Generation', *The Telegraph*, 30 June 2007. http://www.telegraph.co.uk/technology/3353790/The-myth-of-the-MySpace-generation.html (accessed 15 June 2009).

Murdock, G. (1999) 'Rights and Representations: Public Discourse and Cultural Citizenship', in J. Gripsrud (ed.), *Television and Common Knowledge*, London: Routledge.

Negroponte, N. (1995) *Being Digital*, London: Hodder and Stoughton.

Nelson, T. E., Clawson, R. A. and Oxley, Z. M. (1997) 'Media Framing of a Civil Liberties Conflict and Its Effect on Tolerance', *The American Political Science Review*, 91 (3): 567–83.

Newburn, T. (1996) 'Back to the Future? Youth Crime, Youth Justice, and the Rediscovery of "Authoritarian Populism"', in Jane Pilcher and Stephen Wagg (eds), *Thatcher's Children: Politics, Childhood and Society in the 1980s and 1990s*, London: Falmer Press.

Newton, K. (1999) 'Mass Media Effects: Mobilization or Video Malaise', *British Journal of Political Science*, 29(4): 577–99.

Nieminen, H. (2007) 'Disobedient Media – Unruly Citizens: Governmental Communication in Crisis', in Bart Cammaerts and Nico Carpentier (eds), *Reclaiming the Media: Communication Rights and Democratic Media Roles*, Bristol: Intellect Press.

Norris, P. (1996) 'Does Television Erode Social Capital? A Reply to Putnam', *Political Science and Politics*, 29 (3): 474–80.

Norris, P. (2001) 'A Failing Grade? The News Media and Campaign 2000', *Press/Politics*, 6 (2): 3–9.

Norris, P. (2002) *Democratic Phoenix: Reinventing Political Activism*, Cambridge: Cambridge University Press.

OECD (2008) *Growing Unequal? Income Distribution and Poverty in OECD Countries*.

OFCOM (2007a) *New News, Future News. The Challenges for Television after the Digital Switch-over*, London: OFCOM.

OFCOM (2007b) *New News, Future News: Research and Evidence Base*, London: OFCOM.

OFCOM (2007c) *Public Service Broadcasting: Annual Report 2007*, OFCOM.

OFCOM (2008a) *Media Literacy Audit: Report on UK Children's Media Literacy*, London: OFCOM.

OFCOM (2008b) *Media Literacy Audit: Annex 3: Report on UK Children by Platform*, London: OFCOM.

Offe, C. (1985) 'New Social Movements: Challenging the Boundaries of Institutional Politics', *Social Research*, 52 (4): 817–68.

Osgerby, B. (1997) *Youth in Britain since 1945*, Oxford: Blackwell.

Park, A., Roger Jowell, John Curtice, Alison Park, Katarina Thomson, Lindsey Jarvis, Catherine Bromley, Nina Stratford (eds) (2001) *British Social Attitudes: The 18th Report*, London: Sage/NCSR.

Pasek, J., Kenski, K., Romer, D. and Jamieson, K. H. (2006) 'America's Youth and Community Engagement', *Communication Research*, 33 (3): 115–35.
Patterson, T. (1994) *Out of Order*, New York: Alfred A. Knopf.
Patterson, T. E. (2000) 'Doing Well and Doing Good: How Soft News and Critical Journalism Are Shrinking the News Audience and Weakening Democracy – and What News Outlets Can Do about It', Cambridge: Joan Shorenstein Center on the Press, Politics and Public Policy.
Pearson, G. (1983) *Hooligan: A History of Respectable Fears*, London: Macmillan.
Petley, J. (2009) 'What Fourth Estate?' in Micael Bailey (ed.), *Narrating Media History*, London: Routledge.
Pew Research Center (1990) *The Age of Indifference: A Study of Young Americans and How They View the News*, Washington, DC: Times Mirror Center.
Pew Research Centre (2004) 'Cable and Internet Loom Large in Fragmented Political News Universe'. http://people-press.org/reports/display.php3?ReportID=2000 (accessed January 2008).
Pew Research Centre (2007) 'Public Blames Media for Too Much Celebrity Coverage'. http://people-press.org/report/346/public-blames-media-for-too-much-celebrity-coverage (Accessed January 2008).
Philo, G. (1990) *Seeing and Believing: The Influence of Television*, London: Routledge.
Philo, G. (1995) *Glasgow Media Group Reader, Vol. 2 Industry, Economy, War and Politics*, London: Routledge.
Philo, G. (2002) 'Television News and Audience Understanding of War, Conflict and Disaster', *Journalism Studies*, 3 (2): 173–86.
Philo, G. and Berry, M. (2004) *Bad News from Israel*, London: Pluto Press.
Power Report (2006) *Power to the People: An Independent Inquiry into Britain's Democracy*, York: Rowntree Trust.
Prior, M. (2003) 'Any Good News in Soft News? The Impact of Soft News Preference on Political Knowledge', *Political Communication*, 20: 149–71.
Puar, J. and Rai, A. S. (2002) 'Monster, Terrorist, Fag: The War on Terrorism and the Production of Docile Patriots', *Social Text*, 20 (3): 117–48.
Putnam, R. (2000) *Bowling Alone: The Collapse and Revival of American Community*, New York: Touchstone.
Rampton, S. and Stauber, J. (2003) *Weapons of Mass Deception*, London: Constable and Robinson.
Reiner, R. (2002) 'Media Made Criminality: The Representation of Crime in the Mass Media', in Mike Maguire, Rodd Morgan and Robert Reiner (eds), *Oxford Handbook of Criminology*, Oxford: Oxford University Press: 376–416.
Riddell, P. (1998) 'Members and Millbank: The Media and Parliament' in Jean Seaton (ed.), *Politics and the Media: Harlots and Prerogatives at the Turn of the Millennium*, Oxford: Blackwell.
Robinson, P. and Goddard, P. (2006) 'Measuring Media Autonomy in Wartime: An Analysis Of UK TV News Coverage of the 2003 Iraq War', conference paper delivered at International Studies Association Convention, San Diego, 22–25 March 2006.
Rose, Steven (1992) 'Do Animals Have Rights?', *International Socialism Journal* (54).
Russell, A., Fieldhouse, E., Pudam, K. and Kalra, V. (2002) *Voter Engagement and Young People*, The Electoral Commission.
Sabato, L. J. (1991) *Feeding Frenzy: How Attack Journalism Has Transformed American Politics*, New York: Free Press.

Scammell, M. and Semetko, H. A. (1999) 'The News Agenda' in P. Norris, J. Curtice, D. Sanders, M. Scammell and H. A. Semetko (eds), *On Message: Communicating the Campaign*, London: Sage Publications.

Schlesinger, P. and Tumber, H. (1994) *Reporting Crime: The Media Politics of Criminal Justice*, Oxford: Oxford University Press.

Stradling, R. (1977) *The Political Awareness of School Leavers,* London: Hansard Society.

Such, E., Walker, O. and Walker, R. (2005) 'Anti-War Children: Representation of Youth Protests against the Second Iraq War in the British National Press', *Childhood,* 12 (3): 301–26.

Tapscott, D. (1998) *Growing Up Digital: The Rise of the Net Generation,* Maidenhead: McGraw-Hill.

Teeple, G. (2000) *Globalisation and the Decline of Social Reform,* Toronto: Humanity Books.

UK Children Go Online: Final report of key project findings. ESRC and E-Society 2005.

UNICEF (2007) *Child Poverty in Perspective: An Overview of Child Wellbeing in Rich Countries,* Florence: UNICEF Innocenti Research Centre.

Valentino, N., Beckmann, M. and Buhr, T. (2001) 'A Spiral of Cynicism for Some: The Contingent Effects of Campaign News Frames on Participation and Confidence in Government', *Political Communication,* 18 (4): 347–67.

van Zoonen, L. (2005) *Entertaining the Citizen: When Politics and Popular Culture Converge,* Oxford: Rowman and Littlefield.

Walker, A., Kershaw, C. and Nicholas, S. (2006) *British Crime Survey 2005/06,* London: Home Office.

Waterhouse, Keith (2005) 'Hoodies and Goodies in the "Hood"', *Mail,* 16 May, p. 16.

White, C., Bruce, S. and Ritchie, J. (2000) *Young People's Politics,* York: Joseph Rowntree Foundation.

Wilkin, Peter (2001) *The Political Economy of Global Communication,* London: Pluto Press.

Willis, P. (1977) *Learning to Labour: How Working Class Kids Get Working Class Jobs*, Farnborough: Saxon House.

Winston, Brian (2002) 'Towards Tabloidization? Glasgow Revisited, 1975–2001', *Journalism Studies,* 3 (1): 5–20.

Wood, A. M. L. (2008) *Making Our Communities Safer,* Cardiff: Plaid Cymru Policy Development Unit.

Wood, E. M. (1996) *Democracy against Capitalism,* Cambridge: Cambridge University Press.

Wring, D., Henn, M. and Weinstein, M. (1999) 'Young People and Contemporary Politics: Committed Scepticism or Engaged Cynicism?' in Justin Fisher, Philip Cowley, David Denver, Andrew Russell (eds), *British Elections and Parties Review,* 9, London: Frank Cass.

Young, D. and Tisinger, R. (2006) 'Dispelling Late Night Myths: News Consumption among Late-Night Comedy Viewers and the Predictors of Exposure to Various Late-Night Shows', *Journal of Press/Politics,* 11 (3): 113–34.

Index